P9-BYL-717

PRAISE FOR

Dream Cities

"An ambitious study of the forms and ideas of the contemporary city. . . . Graham wants us to see these urban and architectural forms afresh, not as the drab commonplaces they have become but as the work of visionaries 'whose dreamed-of cities became the blueprint for the world we actually live in.'"

—*New York Times Book Review*

"Offers a fascinating look at seven trends in urban planning that shaped the modern world." —*Shelf Awareness*

"An excellent and novel exploration of key ideas behind city spaces and the behaviors they engender. . . . Mr. Graham is as masterly as a novelist when it comes to character development and narrative."

—*Wall Street Journal*

"An intriguing architectural history and an effective antidote to the excesses of urban renewal and city planning." —*Kirkus Reviews*

PRAISE FOR

American Eden

"Mr. Graham recounts his tale with considerable verve and a vast erudition in the history of gardening and the arts generally. . . . Among much else, Mr. Graham shows us that the history of how our nation grew can be found in what it has grown."

—*Wall Street Journal*

"We are what we plant, LA-based writer Wade Graham posits in his history of gardens. When he isn't explaining the economic and cultural influences, he crafts fascinating profiles. . . . An engaging look at our own pieces of paradise." —*Los Angeles Magazine*

"This 459-page collection of landscape design history in this country is enjoyable reading. It is well researched, posing an interesting historic tie from the past to the present."

—*Washington Post*

"Wade Graham gives an informative and absolutely engrossing narrative of how the garden is caught up in the crosscurrents of American history and culture. *American Eden* is an astute analysis—and, ultimately, a joyous celebration—of 400 years of ingenuity and vision. A better or more appropriate book to read in the park or on the deck can hardly be imagined."

—Ross King, author of *Brunelleschi's Dome: How a Renaissance Genius Reinvented Architecture*

"Graham takes a panoramic perspective in his bold interpretation of the form, function, and meaning of American gardens. . . . This blazingly fresh, critical, and ecologically astute masterwork brilliantly traces the great cycles of American life through a spectrum of gardens that embody our devotion to the art of cultivation for beauty and status, sanctuary and sustenance."

—*Booklist* (starred review)

"A fascinating and illuminating tour of this American landscape."

—*Publishers Weekly*

Dream Cities

Also by Wade Graham

American Eden: From Monticello to Central Park to Our Backyards: What Our Gardens Tell Us About Who We Are

Dream Cities

Seven Urban Ideas That Shape the World

Wade Graham

HARPER PERENNIAL

NEW YORK • LONDON • TORONTO • SYDNEY • NEW DELHI • AUCKLAND

A hardcover edition of this book was published in 2016 by HarperCollins Publishers.

FIRST HARPER PERENNIAL PUBLISHED IN 2017.

Designed by Renato Stanisic

Library of Congress Cataloging-in-Publication Data has been applied for.

ISBN 978-0-06-219632-3 (pbk.)

18 19 20 21 OV/LSC 10 9 8 7 6 5 4 3

Contents

Introduction

In order to go forward and consider the city that might be, we must look at the many visions of our cities since the beginning of the massive urbanization that marks this century. What have the proposals been? Have they been tested, and if so, what have we learned from them? What were the values that guided their authors, and to what extent has society itself changed in the unfolding of the saga of twentieth-century urbanism?

—MOSHE SAFDIE, 1997

D*ream Cities* is a book that explores our cities in a new way—as expressions of ideas, often conflicting, about how we should live, work, play, make, buy, and believe. It tells the stories of the real architects and thinkers whose dreamed-of cities became the blueprints for the world we actually live in. From the nineteenth century to today, what began as visionary concepts—sometimes utopian, sometimes outlandish, always controversial—were gradually adopted and constructed on a massive scale, becoming our concrete reality, in cities around the world, from Dubai to Shanghai to London to Los Angeles. Through the lives of these pivotal dreamers and campaigners and those of the acolytes and antagonists who translated or fought their plans, we can trace the careers not just of the urban forms that surround us—the houses, towers, civic centers, condominiums, shopping malls, boulevards, highways, and spaces in between—but also of the ideas that inspired them, embodied in buildings, neighborhoods, and entire cities. *Dream Cities* aims to show how to see the world we already inhabit in a new way, to see where our urban forms—our architectures—come from, how they intend to shape us, how we shape them in turn, and how we participate in the whole process.

Conventional histories of architecture talk about style, and the uniqueness of buildings as though they are art objects, like paintings, independent of their context. But cities are all context, made up not simply of buildings but of assemblies of forms and the spaces and relationships between them, and between this built environment and us. These architectures are what make cities real: each is a design, intention, assertion, thrust, or counterthrust in the long-running battles and debates about how we ought to live. *Dream Cities* is a history of where these forms came from and how they work, and a field guide for identifying and decoding them.

Sometimes we build cities all at once, like new towns, suburbs, or capitals right from the drawing board. Mostly, they are built over time, by different parties, with various bits superimposed on or pushing into one another, displacing, erasing, destroying, replacing, and being replaced. This is why so many cities feel disjointed and complex—they can be read as layers recording change over time as traces, stacks, and collisions, or like partial graveyards of old architectures, many that still live on, being added to, abandoned, or repurposed. Cities can also be read as battlegrounds where different architectures compete, each with an agenda, and economic and political consequences. There are winners and losers. The program of one kind of urban form is often incompatible with other values embodied in other kinds of form. Some use more resources than others, or take resources from another. Some divide and alienate us from other people, from the shared life of communities, or from responsibility for the planet. Always, when a dream architecture becomes a reality, there are unintended consequences.

The glass towers of downtown business districts stand against traditional neighborhoods with their mix of residences, commercial buildings, and community spaces; the leafy streets of affluent gated enclaves seem to deny the concrete cell blocks built to house the masses in so much of the world, cut off from the rest of the city by elevated highways; pseudo-rural, car-dominated, suburban sprawl competes with the density and inclusiveness of pedestrian-scaled neighborhoods, whether in traditional older city centers or in recent mixed-use developments; huge, hyper-designed "experience" shopping malls and ground-up mini-cities built around retail replace Main Street mom-and-pop stores and redefine the boundaries of public and private space in our lives.

Architectures are expressions of the desires of their designers and builders; these forms intend to shape people and thus shape the world. As churches intend to make people pious, prisons to make them obedient, schools to make them attentive, or monuments to make them civic-minded, we build our cities with intent, each type of built form designed to produce a given outcome, provide a particular environment, promote or discourage a certain behavior. *Dream Cities* is concerned with the new architectures of the modern world, as envisioned by their most influential dreamers. Tall towers and fast highways intend to make us feel and act modern and efficient; monumental museum districts, parks, avenues, and plazas to make us feel orderly and civic; shopping malls to make us want to buy, and in doing so to make us happier; sprawling suburbia to make us feel independent and free of the city and its crowding, and when historically themed, to make us forget our present reality by transporting us to other time realms; "eco-friendly" developments to make us feel in enlightened balance with nature.

It would seem logical that urban forms should reflect the enormous differences in climate and geography where people live. The Inuit igloo, the Puebloan cliff dwelling, and the white masonry villages of the Mediterranean are marvelous adaptations to their surroundings. Topography, geology, winds, temperatures, and weather all played a part in how traditional cities were built. But these days, local variation is hard to spot. In the modern era (since about 1850 in Western Europe and America and now everywhere), cities look more alike than they do different, from Singapore to Ulan Bator to Boston to Moscow to Buenos Aires. Aside from those parts of them built before the modern era—the odd churches, squares, and low-rise historic districts—there is a remarkable, global urban monotony: here are tower blocks, there freeways, there shopping malls, over there pseudo-historic suburbs, here a formally ordered civic center, beyond that, mile after mile of car-dependent sprawl. These elements make up the major part of most modern cities, and yet they are rarely mentioned in tourist guides or architectural histories, or even acknowledged.

Each visionary thinker in this book intended his or her architecture as an improvement on the state of things. Many describe their architecture and urban design as a sort of medicine, designed to cure the diseases of the modern city, which, we will hear over and over again, ail us. They often rail against modernity: the enormity, chaos, and crowding of industrial and postindustrial cities. Some call for a retreat to a simpler past, or to the imagined peace of the agrarian countryside, or to basic nature itself. Others look to a future when new technologies such as railroads, automobiles, or computers will liberate us from the dystopian city and bring forth a new, nearly utopian way of living. Concerned with buildings and the spaces in between them, they approach the problem of improving human shortcomings through the improvement of things: buildings, housing, transportation, or technologies. Explicitly or implicitly, the visionaries profiled in this book tend toward embracing what the theologian Reinhold Niebuhr called "the doctrine of salvation by bricks alone." Putting our best efforts into reforming the built environment as the means to reform ourselves and society is a remarkably deeply held belief in our culture, as if we modern urban dwellers are a cargo cult, putting faith in things to transform our souls and spirits.

The results have been uneven. Some of the dream cities made real described in this book have had less than happy consequences—in the destruction of older urban fabrics that we in retrospect recognize for their virtues, and in the displacement of millions of people from familiar homes into unfamiliar, often dysfunctional landscapes. The urge to reform the congested city has meant, all over the world, the enabling and acceleration of car-dominated, resource-gobbling sprawl. Yet other efforts to reengage the city on its own merits while recognizing its challenges, and to reimagine it for an evolving modernity, have been heroic, lifelong campaigns for the visionaries at their forefront.

Dream Cities is not a history of Utopias, though some of the blueprints decoded in it were meant to be utopian. Nor is it a polemic about the failings of our cities; it offers no model for a more perfect world. In the words of the historian and critic Lewis Mumford: "In the end, I promise, I shall make no attempt to present another utopia; it will be enough to survey the foundations upon which others build." *Dream Cities* aims instead to tell the stories behind much of our built environment, to narrate the dreams and intentions—mostly unexamined—behind the now-mundane forms of the modern city. It strives to give the reader the tools to identify the architectures all around us (a field guide, as it were, such as one would use to identify types or species of creatures in nature, and to read, decode, and understand the implications of them); to train ourselves (to continue the analogy with the natural world) to know these urban types by their plumage, their calls, their habitats, and behaviors; and to recognize how they actively shape our lives, while we—most of us, most of the time—go about our daily business.

Dream Cities

Chapter [1]

Castles

Bertram Goodhue and the Romantic City

All history is the history of longing.

—T. J. Jackson Lears, *Rebirth of a Nation*

Safe upon the solid rock the ugly houses stand:
Come and see my shining palace built upon the sand!

—Edna St. Vincent Millay, "Second Fig"

I grew up in sixteenth-century Andalusia. Or so it might have seemed. Most of the small city where I was born is made of white stucco buildings with deep-set windows guarded by wrought iron grates and topped with red roofs. The courthouse is resplendent with brightly colored Tunisian tiles and crowned by a baroque clock tower; even its jail resembles a Moorish palace. The scent of orange blossoms fills the air in winter, and bougainvillea and roses drape over homes and businesses alike. The town's sea of red roofs is presided over by the twin bell towers of an old Spanish church. The street grid is laid out on the 45-degree tilt specified by the Laws of the Indies, the street names drawn from the surnames of the town's first Spanish settlers. A handful of the original adobe structures have been restored and stand as reminders of historical continuity. But 99 percent of the city is fake. This is Santa Barbara, California, 90 miles north of Los Angeles, mostly constructed in the twentieth century and still being built in the twenty-first by Americans flush with industrial fortunes intent on living inside a full-scale stage set of someone else's vanished past.

As a kid, I traversed this faux-Mediterranean idyll on my skateboard. It all seemed perfectly natural. As I got older, I absorbed the mostly unspoken narrative that this ambience and attention to detail were what made us in Santa Barbara different, and, it was implied, better than those condemned to live in LA, the smog- and traffic-choked Babylon to the south. This story was reassuring: the town's antiquated style conferred both a measure of virtue and some degree of protection, because of its separation in time and space from the degraded and degrading Big City.

The separation in space was easy enough to understand: to leave the traffic and smog to others not fortunate enough to live here was evidently a good thing. The other part—the vague sense of virtue that the town's architecture promised—was harder to nail down. Gradually, I came to understand that what was at play was psychological, the achievement of a sense of time travel to a better world. For it to work, the place needed to be more than simply many miles away from the city. It had to be apart from it, as if it occupied a time in the past, a separation more important than physical distance and much harder to bridge. At the simplest level, the veneer of antiquity promises those who invest in it a good return, not just in property value but in worth—in the social-worth and possibly the self-worth sense. A patina of antiquity in real estate can vest a person or family with the aura of old money—which is why the nouveaux riches of all eras have chosen to buy castles to launder their money, washing that bad, arriviste smell out of their wealth. In Santa Barbara the money has mostly always been "new," as each generation of new arrivals brings its loot from some distant capitalist battlefield, retiring from the fray in genteel fashion in this paradise.

The illusion of standing apart in space and time also satisfied another longing: to not live in a modern city at all, to have nothing to do with what the city stands for—work, toil, struggle, urgency, and other people, especially undesirable ones. Santa Barbara is a modern city made successful by pretending that it is neither thing. Its new "old" architecture is an illusion that sustains the collective delusion of difference, which is what makes it so desirable. It forms the basis for a limited-access Utopia.

After a few trips down south as a teenager, I also came to see that the supposed hell of Los Angeles was largely filled with the same stuff: many neighborhoods were lined with Spanish-style houses and commercial buildings, and others were cobbled together from a myriad of different historicist and equally ostentatious modernist styles. The difference was that in sprawling LA there were big gaps in the continuity, as they say in the movie business, so the illusion was rarely as perfect. Yet the basic business these areas were aiming at was the same: re-creating historical architectures as a way of conjuring that golden sense of separation. It was all a sort of a real estate pageant, with romantic trappings that were just that—costumes, veneer, finery, plumage to attract, shiny jewelry to dazzle and distract, ourselves as much as others.

Once you saw what it was, you saw it everywhere: as a part of cities in the form of exclusive outer suburbs or inner enclaves, or sometimes as the basic fabric of whole cities. You can see historical re-creation all over North America, Europe, indeed, all over the world in contemporary cities. The practice began in the nineteenth century when industrial, urban modernity first appeared, then spread through the global "West" in the twentieth century, and continues spreading now, in the twenty-first, as industrial modernity does. It is a curious phenomenon: as we advance, we reach backward in time.

All of it begs the question: Why? Where did it come from? What cultural need made this, and keeps making it? What does it do for us, that we are willing and eager to invest so much in it? How does the magic work? The answer has always been right there in plain sight: in the ponderous white houses along the streets, with their carved oak doors and their gardens of myrtle hedges and lemon trees, or perhaps in the country club up on the hill, crowned by a stout tower like an ancient battlement, where fortunate members play 18 holes of golf on a weekday, overlooking the sparkling Pacific.

The original designer of much of this place was one of the greatest architects to have ever worked in America, and you have most likely never heard of him. You may have seen one or more of his buildings—maybe the Nebraska State Capitol, with its iconic "Sower" sculpture throwing seeds from its 400-foot-tall tower, or the magnificently detailed Art Deco–crossed-with-Mediterranean Los Angeles Central Library, or one of his stunning Gothic-style churches in New York, Boston, or Chicago, but you probably don't associate them with or even recognize their author's name, Bertram Goodhue. Why? Partly because, in the opinion of the modernist critics who came after him and wrote the architectural history books we read today, Goodhue didn't make "modern" buildings, so they consigned him to the dustbin of history, crowded as it is with quaint cornices, columns, pointed arches, and ornaments—what the Viennese proto-modernist architect Adolf Loos famously called "crime." Yet the modernists missed the point. In a paradoxical way, Goodhue, while drawing on the architectural forms of the past, in fact drew the plans for huge swaths of the global contemporary city, precisely by rejecting the forms and spaces of modernity. In their place, he, and many others before and after him, substituted an antimodern, antiurban world of traditional symbols and forms. They performed the magic trick of convincing us to accept the modern world in the moment of privately rejecting it.

It all began with a bit of make-believe.

Bertram Grosvenor Goodhue was born on April 28, 1869, in Pomfret, Connecticut, into the fading glory of a once-illustrious New England Yankee family. He counted five ancestors who had sailed on the *Mayflower* and six who fought in the Revolutionary War. Young Bertram was artistic from the beginning, guided by his mother, Helen, who taught him at home, especially music and art, in two small studios, side by side in the attic. She told him the tales of Saint Francis of Assisi and Saint Augustine, and read him the Arthurian legends and the *Song of Roland*. Early on he showed unusual drawing talent. At the age of 9 he intended to become an architect. He began school at age 11, boarding in New Haven; the other students said that he spent most of his time there "drawing dream cities or caricaturing his fellow students."

With the family's fortunes in some decline, there wasn't money for Bertram to attend Yale, where many of his forebears had studied, or the main avenue toward a career in architecture for wealthy young Americans, the École des Beaux-Arts in Paris. So in 1884, at 15 years of age, he moved to New York City to take an apprenticeship with the firm of Renwick, Aspinwall and Russell, as an office boy, for five dollars a month. He learned fast and soon became a draftsman. He joined the downtown Sketch Club, where he was popular, and was remembered as boyish, blond, with blue eyes and red cheeks, and possessed of a remarkable youthful energy.

After five years, Goodhue was ready to go out on his own. In 1891, he entered and won a design competition for the Cathedral of Saint Matthew in Dallas, in the popular Gothic style—though it was never built. He also entered a competition for New York's Cathedral of Saint John the Divine, with less success, but noticed another entry by the Boston firm Cram and Wentworth, which piqued his interest. That year he went to Boston to meet Ralph Adams Cram, an architect five years his senior who had recently established a partnership with Charles Wentworth, an engineer. Cram offered to share his office space; after a year, Goodhue was brought on as third partner. Cram was poised to become the premier architect of Gothic churches in America, having set his sights on the title after experiencing a quasi conversion during a Catholic mass in Rome. Setting aside his austere New England Unitarianism, he became a devotee of the new aesthetics of the Oxford, or Anglo-Catholic, movement, obsessed with ritual, symbolism, and the Gothic Revival in architecture, begun in England a half century earlier by A. W. N. Pugin and beginning to take hold in the United States. Among the first jobs Goodhue and Cram collaborated on was the Church of All Saints, in Ashmont, Massachusetts, begun in 1891, a battlemented, vaguely Norman Gothic design. They would become the greatest church builders in America in their era: until the dissolution of their partnership in 1914, they built 40 churches and chapels, almost all in Gothic modes, all over the United States, including the masterpieces of the Chapel at the US Military Academy in West Point, New York; the Rockefeller Chapel in Chicago; and in New York City alone; the Chapel of the

Intercession, Saint Thomas Church, Saint Bartholomew's Church, the Dutch Reformed ("South") Church, and the Church of Saint Vincent Ferrer.

The firm's presentation drawings were done by Goodhue in deft, dense, confident strokes of pen or graphite, sometimes rendered with watercolors. They exude an atmosphere that is both calm and familiar and somehow mysterious and exotic, like windows onto other worlds. Even amid the weight of the verisimilitude there is often a hint of whimsy in the drawings, an inflection in a detail or vignette that injects humor into the enterprise. Cram articulated this quality when he recalled of his partner: "His pen and ink renderings were the wonder and the admiration of the whole profession, while he had a creative imagination, exquisite in the beauty of its manifestations, sometimes elflike in its fantasy, that actually left one breathless."

Between 1896 and 1899, then in his late twenties, Goodhue penned a series of detailed traveler's reports of places visited on a tour of Europe. He wrote about three romantic, out-of-the-way, overlooked sites that still boasted ancient examples of buildings and bygone patterns of life. These sites he rendered in exquisite ink drawings: careful architectural plans of major buildings, and perspective views showing quarters of towns or groupings of buildings, or their situation in the countryside or gardens, and everywhere including scenes depicting the daily life of the inhabitants. Lively notes recalling the architect's visits and his conversations with the people accompanied the drawings. The first portfolio, done in 1896, was *Traumburg*, which portrays a medieval village in German Bohemia, dominated by a Gothic cathedral-size Saint Kavin's Church, presented in careful plan form, with columns and vaulting ribs in a specific pattern seen elsewhere only at Ely Cathedral, built in the fourteenth century in Norfolk, England. The church fronted on Kavinsplatz, a public square, shown by Goodhue in perspective drawings filled with loving detail of townsfolk walking and peasants driving horse-drawn carts. Down a lane a maiden stands under a baroque bay window, gazing at a distant sentry under an arched passage. Above a steep, shingled rooftop, storks sit on a twig nest perched atop a brick chimney; the church's Gothic spire rises in the distance. From across the river outside the town, looking over a stone bridge, the half-timbered houses cluster around the massive, monumental church, the whole town like a coral clinging to a rocky outcrop. The tower, with its incredibly intricate ornament, is of extraordinary size, both massive and tall, looming up over everything around it.

The second, done in 1897, was *The Villa Fosca and Its Garden*, which depicts a Renaissance villa on a remote island in the Adriatic Sea. It is first seen in an oblique perspective approaching the entrance—a forecourt wrapped on three sides by the two-story Italian building with tile roofs. A plan of the first-floor rooms includes their adjacent formal garden features, including the "Grotto of Hecate, Fountain of the Satyrs, Exhedra with Group of Three Dancing Figures (the Graces?), and Statue of Silence." The villa, Goodhue observed in his notes, "has fallen quietly from its once high estate into a present condition of hidalgo-like decay." Viewed from the garden side, partly reflected in a broad pool below a massive staircase leading up to the arched and columned Roman facade, the villa seems to him overwrought, revealing "the vaulting ambition of its designer, apparently some dry-as-dust pupil of Vignola," the sixteenth-century Italian mannerist.

The third portfolio, from 1899, rendered Monteventoso, a village in northern Italy notable for its Church of Santa Caterina and central Piazza Re Umberto. From across the neighboring valley, a viewer in a streamside meadow could make out the hill town's timeless form: houses with a blocky, vaguely Spanish feel begin at the edge of the fields, then cluster more and more thickly up the lower slopes of a mountain, becoming a dense jumble around the cathedral, with its three-arched Gothic facade, topped by a colonnaded dome like Saint Paul's Cathedral in London, and a square campanile reaching proudly skyward behind it. A perspective of Piazza Re Umberto depicts life in the plaza: people shopping or strolling, a lady selling vegetables under an umbrella in her market stand and waiting for customers. Like the drawings, Goodhue's meticulous notes captured the animated scene: "its musical and unmusical sounds, its clamouring people, its miserable bronze Umberto, and rickety iron café tables." He recorded a long conversation about art and music with one townsman, and, in language as tactile and evocative as his contemporary Henry James, the feel of the place at a day's end:

> Below me in the now windless and shimmering atmosphere huddled the purple and red roofs of the town, the torturous streets marked by narrow courses of liquid purple through the gold and salmon roofs and walls. From the midst of all this color rose the campanile, clear-cut against the hazy distance, the detonation of its bells on the instant breaking the air into an invisible tempest, while its forked battlements seemed less to bring to mind "old, unhappy, far-off things, and battles long ago" than to accent the peace and stillness of today, the time and place.

Each report had a good dose of architectural and historical scholarship behind it—and the author's occasional disapproval: "The buildings are all flawed!" he wrote, though his remarks were couched in the light tones of typical late Victorian tourist writing. The reports were guides to using architecture as a means to creating experience, to setting a scene—complete, inhabited, synesthetic with color, sound, and language, and happening before our very eyes—and setting it so well as to transport readers there to almost experience it themselves. Primers in the art of staging a perfect illusion—and rightly so, as Goodhue had at that time never been to Europe—they were pure romance. Traumburg could be translated from German as "dream town" or perhaps "dreamville," Monteventoso as "windy mountain," and Villa Fosca as "Gloom House." The reports fell into the long tradition of *voyages imaginaires*, like Thomas More's *Utopia* (1516) or Francis Bacon's *New Atlantis* (1627), and were equally utopian—in the design sense, if not the political one. The difference was, Goodhue was not a philosopher, but an architect, and these reports weren't idle romantic or artistic exercises, but careful studies, the first of many over a long career, of how to make dream cities real.

The decade of the 1890s was the period of Cram and Goodhue's closest partnership and artistic ferment—stoked without doubt by their enmeshment in the young bohemian milieu of Boston and Cambridge, where they were members of boisterous student drinking clubs like the Pewter Mugs and of avant-garde arts collectives like one called the Visionists and the Boston Art Students' Association. The latter group staged romantic plays; a picture of Goodhue in costume survives, showing him looking exuberant in a false mustache. In 1897, Cram and Goodhue helped found the Society of Arts and Crafts in Boston, dedicated to the revival of traditional crafts and design and fascinated by all things medieval. Inspired by the example of the English Arts and Crafts movement leader William Morris's hand-lettered and -illustrated edition of the *Works of Geoffrey Chaucer*, which Morris's Kelmscott Press published in 1896, Goodhue produced *The Altar Book of the Episcopal Church*, an exquisite illuminated volume every bit the equal of Morris's masterpiece. With Cram and other friends, he worked on a short-lived review, the *Knight-Errant*, drawing the covers in pen and ink, including one depicting a mounted knight in armor in a valley brook, gazing up at a castle on the hill above.

Goodhue's talent and effort were prodigious: in addition to architecture and the arts, he found time to create a typeface, named Cheltenham, which remains in use. Prodigious also was his vitality, which energized the Boston group in the years 1890–1900. "He more than anyone else, was instrumental," Cram later wrote. "The sense of romance possessed him . . . and made him into a Medievalist in all things." Goodhue loved role-playing, play-acting, and music—all the elements of theater. He affected a romantic style as if he were always in character: "perched on a table, smoking, a broad hat slouched over his eye, a cigarette smoldering under his blond mustache, a Mexican 'capa' flung over his shoulder while he strummed out improvised accompaniments on a battered old guitar."

The name Arts and Crafts sounds quaint today, and irrelevant, something to do with harmless habits of homemade pottery and stained glass. But during its more than half-century run it was a proper movement, set on changing the world, among the most wide-ranging, ubiquitous, energetic, and influential artistic movements in modern history. It is also still very much with us, though unacknowledged—having become deeply embedded in our attitudes about value, authenticity, and unique authorship, and having informed its seeming opposite, modernism, so fundamentally that the fact that the latter is the daughter of the former is rarely discussed. The movement wasn't named until the 1880s, by its greatest proponent and practitioner, the British poet, architect, and textile and furniture designer William Morris, but it is conventionally said to derive from the earlier work of the English art critic John Ruskin, whose 1853 book *The Stones of Venice* fixed its worldview and agenda. Ruskin railed against the modern industrial system of production, arguing that machine-made objects alienated the maker from the products of his labor and robbed those who used them of the dignities of work and art made by the human hand. He denounced the Renaissance, with its rationality and commercial values, as a "false dawn," when in the fact the medieval Dark Ages had been the true golden age, with communities of skilled craftsmen forging an organic unity of the laborer, the object, and its user, and through this unity, a larger unity of the land, the people, the nation, the church, and God. He lauded the making of the Gothic cathedrals, some of which took generations to complete, as examples of a moral aesthetics—a way of reintegrating spirituality with the everyday through art. For Ruskin, the correct making of art was the key and kernel of an earthly Utopia; it must express the full range of human capacity, feeling, memory, and meaning, and so return society to its rightful moral and spiritual balance.

Morris, inspired by Ruskin and Pugin and the artists of the Pre-Raphaelite movement, pioneered the resurrection of traditional craft techniques, working in almost every medium in an effort to lift decorative art from a mainly commercial activity to a fine art. He made an astonishing range of stuff: hand-printed books, printed textiles, furniture, jewelry, stained glass, metalwork, and ceramics, replete with images and natural patterns of tendrils, flowers, trees, and birds, and suffused with the ideas of "organic" design, production, and use. Morris established a series of workshops in England from 1861 onward and trained many other workers, some of whom went on to start their own workshops and publications, and the Arts and Crafts movement blossomed, spreading through Britain, the Continent, America, Australia, and elsewhere.

The same reform impulse applied itself to the problem of the city. The squalor of the new industrial cities began to be acknowledged and publicized by writers like Friedrich Engels, who described the horrors of the working-class slums of Manchester in the 1840s, and Charles Dickens, whose novels of London gave the adjective "Dickensian" to this new phenomenon of industrial urban immiseration. "People"—that is, the upper and middle classes—were appalled, and calls to "do something about it" added an urban component to the era's expanding agenda of social reform, which had begun with the antislavery movement and grew to embrace temperance, the improvement of industrial working conditions, wages and safety, child labor, women's suffrage, hygiene, municipal services, good governance, and housing standards.

The overwhelming fact about the built environment in America, Britain, and much of the Continent was that industrial cities grew like gourds in the night while the countryside was gutted as peasants and small farmers were pushed off the land by collapsing prices for farm products. But in culture, things seemed to move in the opposite direction. It might be said that nineteenth-century culture was a constant attack on the very idea of cities—as unnatural, threatening, uncontrollable breeding grounds of vice, disease, danger, and corruption. Moralism saturated the conversation: the city is bad; its opposite is the country, the site of virtue made up of honesty and simplicity, in working directly on the land, in direct, pious feeling, and in simple pleasures. Salvation was to be found in a past golden age—forget the fact that the real countryside bore no resemblance to the mythic one. Even as millions were uprooted from hard agricultural lives and sucked into the vortex of wage labor and poverty in cities, the dominant cultural trope was arcadianism and pastoralism: fleeing the city to a rural, agrarian idyll. Never mind that few could afford it.

But for social theorists of all stripes and architects in particular, the shining path lay not in retreating to the farm but in fashioning a new kind of place: one with the economic and social advantages of the city but none of its disadvantages. It would have just enough urban structure but plenty of access to green space and fresh air, somewhere between the city and the frontier, between the factory and the fields. Utopian thought was rampant and tended to center on the belief that reforming the built environment would reform society and people. A long tradition of socialist idealism linked up with the new aesthetic moralism—exemplified by Ruskin's demand for a "morality of architecture," generating an impressive range of solutions.

For some philanthropically minded industrialists the answer lay in building factories in the fields. Robert Owen built his utopian mill town, New Lanark, in Scotland, in 1799, then tried again at New Harmony, Indiana (1825), and inspired a long line of imitators in many countries. Other industrialists built more paternalistic company towns, from Lowell, Massachusetts, on to Pullman, Illinois (1880), where George Pullman's railroad cars were built, and Port Sunlight, Merseyside, near Manchester, England (1888), built by the soap magnates the Lever brothers and named after their company's cleaning product, Sunlight. All were built in historicist architectural modes, generally Gothic or Tudor, to underline their distance from the soul-sapping modern city.

In the United States, experiments were everywhere. In the 1830s, Alexis de Tocqueville had expressed Americans' belief that they were making the world anew, including its urban models: "The new society . . . has no prototype anywhere." In 1840, Ralph Waldo Emerson wrote to his British friend Thomas Carlyle: "We are all a little wild here with numberless projects of social reform. Not a reading man but has a draft of a new community in his waistcoat pocket." Many experiments were utopian in intent, whether socialist-industrial, or irrigated farm colonies, or artists' colonies, such as the many Arts and Crafts–inspired examples: Byrdcliffe Colony, Brook Farm, Oneida, Modern Times, Harmonia, and Celeste top a lengthy list (the lists from Britain, Ireland, Australia, the Continent, and elsewhere are just as long). In the century from 1820 to 1920, there were more than 250 utopian communities in the United States, the average duration of which was fewer than four years. Many others were proposed: especially new towns to be set up on green land—religious communities, irrigation colonies, or experimental town plans meant to ward off the evils of the city while profiting from the economies that true rural areas lacked. Some put faith in formal innovations such as distinctive geometries: Circleville, Ohio, and Octagon City, Kansas, are just two examples. Others proposed more sophisticated planning, with tight control over land uses, separating zones of different activities from residential to industrial to agricultural, often in layouts of concentric rings. The most influential of these was the "garden city" concept first popularized by the English stenographer Ebenezer Howard in 1898. It called for the building in rural areas of discrete new towns that would combine industry, farming, and small-town scale by strict zone separation and population limits, each an arrangement of concentric circles, buffered from one another and from outside influences by greenbelts. The garden city idea inspired countless twentieth-century planners and more than a few actual attempts—Letchworth, England (from 1905), and Greenbelt, Maryland (from 1935), notable among them. Most of the examples quickly became dormitory suburbs of nearby metropolises, devoid of the integrated industry that was to have made the garden city different. But most such schemes passed into obscurity, as one author put it, "like so many paper soldiers."

And yet, a new, and real enough, sort of Utopia did emphatically come into being around the established cities, without challenging their dominance: the railroad suburb. Passenger railroads radiating out from central cities created a new possibility: working inside, in what became known as central business districts, but living outside, removed, in a pseudo-pastoral Eden out of reach of the poor, the criminal, the immigrant, and their attendant unpleasantnesses. The suburb was a kind of anticity, both conjured into existence by and completely dependent on its umbilical cord to the city—that ultimate industrial technology, the iron horse. So the romantic suburbs were born: first in England in the 1840s, around London, Liverpool, Manchester, and other cities, then in the United States, where the first was Llewellyn Park, New Jersey, a pastorally landscaped residential subdivision complete with a medieval-themed gatehouse, developed by a visionary businessman when a new railroad offered direct service to Manhattan, just 13 miles away. The American landscape architect Frederick Law Olmsted would go on to apply the romantic vocabulary to Central Park in New York. His firm, Olmsted and Vaux, designed 16 leafy suburbs, including Riverside, Illinois; Brookline and Chestnut Hill, Massachusetts; Roland Park, Maryland; and Yonkers and Tarrytown Heights, New York. Developers on the outskirts of industrial cities around the world would follow suit. The combination of relatively cheap land and fast transport (for those who could afford it) was irresistible, and suburbs sprouted around Victorian cities like mushrooms in a field after the rain.

They were not towns, where one might find services and employment, but white-collar bedroom communities, often gated, separated off within bubbles of romantic, picturesque landscaping and medievalist architecture. (Romanticism, as the German sociologist Georg Simmel noted at the time, is an extension of urban sensibilities, a precursor to tourism, and requires money to indulge in.) Though pseudo-rural looking, they had nothing to do with rural life; they were consummately modern and urban phenomena, stage sets inside which people could live. First for the upper class, then the expanding white-collar middle class, it became a social and hygienic imperative to leave the city, while, of course, continuing to make one's living in it. The city and the proper home were now considered "despite their continuing mutual reinforcement, not simply as distinct but as fundamentally incompatible. . . . The city inflicted injury that the home healed." If the city was masculine and all about machines and work and danger, ideal Victorian homes were the opposite: removed in safe "country" places, and exaggeratedly feminine and domestic, full of lace curtains and carefully tended pots of geraniums—ubiquitous symbols of fragile femininity—on the windowsill. Women ruled, but were also dependent—to show that their husbands had no financial need for them to work—and isolated in the home castle, well defended against the threats of the outside, modern world. It was "smokestacks versus geraniums," as one San Diego, California, mayoral candidate's campaign put it in 1917.

Born with the railroad suburb was the commuter, and almost as soon as he appeared he became an archetype—first captured by

Dickens in *Great Expectations* (1861). Mr. Wemmick, clerk for a criminal lawyer in London, handler of cash and the grubby daily details of his criminal clientele in Newgate Prison, is "a dry man, rather short in stature, with a square wooden face, whose expression seemed to have been imperfectly chipped out with a dull-edged chisel." He returns nightly to his little suburban house in Walworth, south of the City: "a little wooden cottage in the midst of plots of garden," the top of which he had "cut out and painted like a battery mounted with guns," and which he called his Castle. He fitted it out with all the accoutrements of a romantic country estate, including a little moat and drawbridge, twisting paths, a garden, an ornamental lake, a fountain, pointed Gothic windows, and a cannon he fired nightly called the Stinger. The change he experienced during his transport between work and home is fundamental, as Wemmick tells his guest, Pip: "The office is one thing and private life is another. When I go into the office, I leave the Castle behind me, and when I come into the Castle, I leave the office behind me." The transformation is almost magical, or pathological, like Dr. Jekyll and Mr. Hyde. As the two men return on the train into the city, Pip observes: "By degrees, Wemmick got dryer and harder as we went along, and his mouth tightened into a post-office again. At last, when we got to his place of business and he pulled out his key from his coat-collar, he looked as unconscious of his Walworth property as if the Castle and the drawbridge and the arbor and the lake and the fountain . . . had all been blown into space together by the last discharge of the Stinger."

Of course the Castle is a fantasy, in which the commuter is a country gentleman instead of a city functionary. But the fantasy is what makes being a city functionary tolerable. In the city he is constantly under pressure of time—being on time, finishing on time, not missing the schedule—so his retreat into an imaginary golden age where time doesn't pass is both a consolation and a form of resistance. The physical separation between work and home mirrors the psychological separation he needs and builds, and these together are what allow Mr. Wemmick and countless others to mask the modernity they live in. It's theater: the dress-up and staging that enfold the Castle and its absurd historical dramas are fantasy; and the city and the world of work are no less a site of fantasy, complete with costume-uniforms, conventions, and necessary delusions about upward mobility, success, fairness, and transcending one's lot. Virginia Woolf, in her 1927 essay "Street Haunting," captured this transformation in the space of the commuter train:

> They are wrapt, in this short passage from work to home, in some narcotic dream, now that they are free from the desk, and have the fresh air on their cheeks. They put on those bright clothes which they must hang up and lock the key upon all the rest of the day, and are great cricketers, famous actresses, soldiers who have saved their country at the hour of need. Dreaming, gesticulating, often muttering a few words aloud, they sweep over the Strand and across Waterloo Bridge whence they will be slung in long rattling trains, to some prim little villa in Barnes or Surbiton where the sight of the clock in the hall and the smell of supper in the basement puncture the dream.

It was no coincidence that this period was also the golden age of children's literature, especially what is called the Edwardian period: from the death of Queen Victoria in 1901 and the accession of Edward VII to his death in 1910—though its end is better marked by the assassination of Archduke Ferdinand in Sarajevo in 1914 and the beginning of World War I. A raft of classics were written in one decade, enough to constitute a new genre of children's romances and an instant canon—neither of which has yet been superseded: Beatrix Potter's *Peter Rabbit* (1902), J. M. Barrie's *Peter Pan* (1904), Edith Nesbit's *The Railway Children* (1905), Kenneth Grahame's *The Wind in the Willows* (1908), Lucy Maud Montgomery's *Anne of Green Gables* (1908), Frances Hodgson Burnett's *The Secret Garden* (1910), and so on. They are all tales about blurring the lines between fantasy and everyday life, set in dilapidated old houses or gardens or literal pastures (the pastoral), and involving magic, adventures, and dressing up—often enough featuring small furry animals dressed as people. Like nineteenth-century culture at large, these tales are bathed in nostalgia for an imaginary rural past, and a fascination with childhood innocence confronting the dangers of a willful, arbitrary, threatening adult world. They were hugely successful, and by no means just with children: in 1904, the biggest hit on the London stage was *Peter Pan, or, the Boy Who Wouldn't Grow Up*.

It was also no coincidence that children's stories flourished when both the United States and the United Kingdom were ruled by men "widely perceived to have never grown up," in the words of children's literature scholar Seth Lerer: Theodore Roosevelt and King Edward VII, men who loved adventurous exploits, dressing up in costumes, and the occasional "splendid little war." Set against people's real-life, day-to-day experience of breakneck social and economic change as well as constant military conflict in the imperial borderlands of the United States and the European powers, these "cartographies of nostalgia," in Lerer's phrase (think of all the maps: the Hundred

Aker Wood in *Winnie-the-Pooh*, Neverland in *Peter Pan*, etc.), were powerfully alluring. In response to a threatening present and future, these stories often took the form of a journey back in time; since the medieval period was cast as the childhood of European civilization, so children's literature, no less than the other major aesthetic movements were for a century, was often set in quasi-medieval milieux. But Peter Pan's Neverland was far less escapist than King George IV's Royal Pavilion at Brighton—a fantasy castle in the invented "Indo-Saracenic" style, a pastiche of Orientalist tropes that was the colorful, gay alternative to gray Gothic; it became the official style of the British occupation of India, and inspired square miles of bad architecture worldwide, from wedding cake London hotels to the American circus magnate P. T Barnum's onion-domed "Iranistan" mansion in Connecticut.

Ultimately, the revolutionary intent of the Arts and Crafts movement failed: handmade objects were too expensive for any but the wealthy, most of whom had gotten rich from the industrialization and standardization of production that the movement decried. In the end, the movement devolved into something essentially conservative, and decorative in the worst sense of the word; thus its architecture and urban creations served to buttress the status quo, not undermine it.

The Arts and Crafts reformers were looking at the wrong things: things. Just as Morris wanted to believe in the power of better-designed and more humanely made objects to cure our social ills, garden city makers and other critics of the conditions of the new industrial city wanted to believe that the city itself was the culprit, not the economic conditions that drove its growth. They failed to look at themselves behind the curtain, like the Wizard of Oz, operating the machinery.

The social problem doesn't reside in things, but in unequal access to power, wealth, education, and resources. Better design will not save us from ourselves.

Though he hadn't visited Europe before composing his three dream-cities travelogues, Goodhue began traveling in his Boston years and never stopped. After fruitless meetings with the sponsors of the Dallas cathedral, he journeyed into nearby Mexico for several months in 1891, traveling by train and occasionally horseback. The next year he published *Mexican Memories*: the record of a slight sojourn below the yellow Rio Grande, a witty, self-deprecating account that is less travelogue than fin de siècle meditation on time and progress. "So far, Mexico has not sunk to the level of other lands," he wrote, "and in this, perhaps, lies its chief charm." The sinking is progress: the relentless speed-up by machines that everywhere eats away at the few places that remain where Western tourists can experience the exotic long ago and far away in real time—the nostalgia in the present moment that is at the heart of romanticism. "Nowa-days, the magic carpet of the Eastern story is left far behind in the march of progress. It is a terrible thing to contemplate, but the vulgar and matter-of-fact tourist can go from Kalamazoo to, say, Nijni Novgorod in no time at all, speaking comparatively, and with infinitely more luxury and ease than any carpet, no matter how 'magic,' could ever afford."

He laments the spread of railroads, but is happy enough to board them after several days of riding leave him sore. Rhetorically, he smoothes over this contradiction by comparing the machines to animals—zoomorphicizing them just as an Edwardian children's author might: "Railroads, in spite of my recent remarks, are, after all, good creatures, and a donkey, or even the beautiful and thoroughbred native horse, becomes a wearisome seat after two or three days' riding." The speed and convenience of the train of course are what made the Orientalist "Eastern story" accessible to mass tourism beginning in the nineteenth century, and they opened to Goodhue a steady progress of romantic prospects. Approaching Mexico City, he described a rhapsodic scene, unfolding as if he had scripted it himself: "Rapidly growing larger and larger as the train approached, was the City itself, whose thousand domes and white walls glistened in the first rays of sunlight like a dream city. Dominating all appeared the towers of the cathedral itself. . . . Tremendous as it is . . . almost overhanging the whole scene."

He took to horseback whenever possible, visiting the country's "delightfully dormant" towns and countryside, admiring the white walls, buildings transported out of medieval Iberia, Indian villages, all the while dressed up as an elegant and armed caballero, sporting "sombrero, zarape, a cool loose shirt, short jacket, silver-ornamented and excessively tight breeches . . . a leathern belt covered with droll Indian stamped work, and filled with cartridges for a certain instrument, the ivory handle of which projects unassumingly." He recounted tales told him in turn, of Old Mexico, always touching on the passage of time—like that of a monk, Frey Antonio, who disappeared from his monastery without a trace, only to reappear 200 years later, baffling himself and the monks occupying his old rooms. He described bullfights, Indians, banditos, señoritas, volcanos, churches, and music. He illustrated his pages with watercolor sketches of towers, seen from an oblique angle, while he was looking up. Everything swelled with enormous romance, imparted in no small part by an urgent nostalgia for something about to disappear. "But you must make haste, though, my friend," he wrote, "for the times are changing very quickly in Mexico and with them the people."

In late September 1894, Goodhue traveled to Quebec and made a series of pencil sketches of the old city, rendered with closely observed detail. They are typically evocative: one shows a square, with children in the foreground facing the viewer, a dog foraging among boxes, carts, and the bustle of the marketplace; another depicts smoke from chimneys, laundry strung across narrow alleys between houses, chickens, and a covered walk. A cathedral on a hill is seen across a broad river dappled with boats and ripples, like a Turner sketch.

Though he may have been the most accomplished architect of fantastically detailed buildings of his time, Goodhue was clearly interested in more than such structures. He was a careful student of the totality of city life: his pictures, in lines and words, subordinated architecture to urbanism, the buildings being important not as objects but as pieces of a bigger, moving drama. Goodhue had read and admired *Architecture, Mysticism and Myth*, by the English Gothic architect W. R. Lethaby, who emphatically put the medieval cathedral at the center of community life, not as architecture but as place: "It would be a mistake to try and define it in terms of form alone; it embodied a spirit, an aspiration, an age." Goodhue echoed him closely: "In France the cathedral stood . . . for the center of everything—religious, governmental, state and civic, as well as personal, so it may almost invariably be found facing the market place with chattering women sitting under umbrellas on its very entrance steps and even with little shops and booths tucked away between its projecting buttresses." The cathedral was the anchor of a miniature city. Hence, Goodhue's drawings contained vignettes of urban worlds. For his and Cram's proposal for the Cathedral of the Incarnation in Baltimore, Maryland, he populated the church's exquisite interiors and exteriors with congregants and choristers going about their business. He drew an extraordinary bird's-eye view of the city, from the ship-filled harbor across from the houses, buildings, streets, and railroad tracks, to the countryside beyond, with the inscription: "A View of the Cathedral City of Baltimore. Behold! The cathedral rises above the city of Baltimore like a finger pointing to the sky. The houses cluster around it like children holding their mother's skirts. Into it are built the faith and prayer of thousands."

Cram and Wentworth was busy throughout the 1890s, building campuses, mostly Gothic churches, and houses in an eclectic range of styles. In 1897, partner Charles Wentworth died, and Frank Ferguson, an engineer in the office, was promoted, so that the firm became Cram, Goodhue and Ferguson the next year. The work was far-flung and entailed grueling hours and much travel, and Goodhue came down with pneumonia. In December 1898, he returned to the dry air of Mexico for four months, this time accompanying the architectural writer Sylvester Baxter and the photographer Henry Peabody, helping them to gather material and prepare drawings for a mammoth study: *Spanish Colonial Architecture in Mexico*. As ever, he paid close attention to the details. It would pay dividends.

In 1901, Goodhue accompanied a new client, James Waldron Gillespie, of New York, a wealthy bachelor and art collector, first to Europe, then on to the Middle East. They traveled through Italy, the Levant, and Persia, where they rode 800 miles on horseback from the Caspian Sea south to the Persian Gulf, passing through Isfahan, Kum, Shiraz, and Tehran. It was truly a voyage into the past, unmarred by the intrusion of railroads. The two men drank in the architecture and gardens of this prelapsarian Oriental dream world, often walking through ancient irrigated walled gardens lit by moonlight. Goodhue made pen-and-ink drawings languid with romance: dark cypress allées, terraces, arched arcades, porticoes, and hidden gardens, all reflected in limpid pools.

On their return, Gillespie commissioned Goodhue to design a house for 30 acres he had bought in Montecito, California, on the edge of Santa Barbara, in the general theme of a Roman villa. What Goodhue produced was extraordinary. On the outside, roman columns embedded in a simplified facade, with a series of bas reliefs of the British Arthurian legends that had been dear to him as a child. The house enclosed a four-sided courtyard and further opened to a stone terrace facing southward to the Pacific Ocean, from which a procession of Persian-inspired shallow pools dropped down a series of terraces connected by stairs into a rectangular pool divided in four sections by walks to a central circular fountain. Another longer and grander staircase flanked by cypresses continued farther down to a series of three long, rectangular pools, pouring into one another and terminating in a colonnaded pavilion. The gardens were studded with the rare palms Gillespie collected, adding to the effect of exotic grandeur, at once highly structured and lushly overflowing its geometric bounds.

Finished in 1906 and named El Fureidis, Arabic for "paradise or haven," it was arguably the first "Mediterranean" house in California, a combination of elements from Persia, Italy, and Spain, but here drawn together in an entirely new form, both historicist and paradoxically modern. "It is not eclectic, it is synthetic," is how a later building of Goodhue's, the Nebraska State Capitol of 1922–32, was described by one essayist, a description that fits El Fureidis equally well. It, too, is a mixture, with Egyptian motifs as well as classical ones, all readily identifiable but even in sum insufficient to explain the structure's essence, which was "rather as a glamour than a memory. The courts and east and west facades speak of Renaissance Italy, but in no sense as an echo, rather as a new voicing of an unforgotten charm."

Glamour and charm distilled from the past were at the heart of El Fureidis's allure, and they came no less from its architecture than from the mood conjured by its setting, its gardens, and its narrative quality—decanted from the swirl of late-Victorian fantasies about Saracens, moguls, sheiks, pharaohs, and knights. It was a real-life Xanadu, as in Samuel Taylor Coleridge's famous poem about Kublai Khan's pleasure dome and walled garden with the river flowing through its center. The poem was one of the most popular of the nineteenth century, and, as if to prove that its imagery wasn't far from his mind, Goodhue made a drawing of a colossal Orientalist castle topped by an enormous dome, reflected in a mountain lake, which he signed "In Xanadu," as a gift for Elmer Grey, a California architect whom he employed to oversee the construction of the house.

This romantic allure was amplified by the new California culture being constructed around it, and, as it happened, inside it. The house was photographed and widely publicized, becoming the prototype for a new, more elegant "Mediterranean" architecture being laid on top of the base of the heavy, low Mission Revival style popularized by California boosters and developers around the turn of the century. It was a perfect stage set; a silent costume drama was filmed there in 1915.

The link between this architecture and Hollywood wasn't accidental. The first filmmakers came to the area in 1908, lured by the clement weather, the variety of scenery for filming, the cheap real estate, and the freedom it offered from Eastern lawyers enforcing patent claims against those using processes claimed by others. They loved the built-in exoticism of the region's Spanish and Mexican past. Moviemakers picked imagined Spanish and Mexican stories as subjects and used several Moorish-style buildings, among the first structures in Hollywood, as sets. In 1908, Colonel William Selig shot several films in Los Angeles, including *Carmen*, in a Spanish courtyard, the first purpose-built movie set in the area. As an actor in California, D. W. Griffith played the role of Alessandro the Indian in a touring play of *Ramona*, based on the best-selling 1884 novel by Helen Hunt Jackson, which had focused national attention on Southern California's romantic past. When Griffith turned himself into a director, one of his first films was staged at the Mission San Gabriel near Los Angeles: *The Thread of Destiny*, with Mary Pickford, about Mexican California. He continued the theme with the titles *In Old California* and his own version of *Ramona*, also with Pickford.

Real estate developers and the Los Angeles Chamber of Commerce from the beginning latched on to the Spanish mythos of arcadian pastoralism to sell their products. The impresario John McGroarty put on a long-running pageant he called *The Mission Play*, also at the partially restored Mission San Gabriel, with the slogan "The happiest land the world has ever known"—sponsored by none other than the streetcar and real estate tycoon Henry Huntington. Hollywood, architecture, and the development industry all fed at the same trough.

To keep pace with the escalating glamour and sophistication of Hollywood films and the publicity culture that fed it, architecture had to evolve—the Mission and Moorish styles soon seemed outmoded and crude. Here is where Goodhue's talent, preparation, and temperament proved ideal. His knowledge of Latin America had deepened with new commissions. In 1905, he designed La Santisima

Trinidad (Holy Trinity) Episcopal Pro-Cathedral, in Havana, Cuba, with a facade and an impressive bell tower encrusted with "churrigueresque" ornament, named for the sixteenth-century baroque style that Goodhue had studied in Mexico on his earlier trips. In 1911, he was hired to design the New Washington Hotel, in Colón, Panama—personally chosen by US president William Howard Taft, who wanted a big statement American hotel on the Atlantic side of the Canal Zone. The same year, he was hired to design the Panama-California Exposition in San Diego, planned by that city's burghers to coincide with the 1915 opening of the Panama Canal, aimed at luring business and immigrants to the little city of 35,000 to the south of Los Angeles. Goodhue responded with a magisterial, fantastical stage set on the chosen site in Balboa Park, a realization of all the dream cities he had so lovingly conceived of over the years. Reached by crossing a long, arched concrete bridge over a ravine, two intersecting streets are lined with Spanish-Mediterranean buildings and threaded with gardens, palms, and fountains. Facing the central plaza is the California State Building, built in the form of a cathedral, crowned by a magnificent blue-tiled dome and a campanile tower encrusted with churrigueresque detail. It was then (and remains today) a world apart, its magic sustained no less by tight control of vehicular and pedestrian access and circulation than by its architectural detail. People loved it. And the architect's San Diego patrons were pleased, as the city was on the map and would go on to attract its share of the investment pouring into the Golden State through new industry, agriculture, immigrants, and military bases bankrolled by the government in Washington, DC.

The renowned astronomer George Ellery Hale summed up the breathless admiration the mini-city in Balboa Park struck in visitors:

> This superb creation, so Spanish in feeling—yet so rarely equalled in Spain—with its stately approach, its walls springing from the hillside, its welcoming gateway, its soaring tower, and its resplendent dome, foretelling all the southern privacy and charm of the courts that lie beyond, reveals much of its author. His constructive imagination, overflowing with visions of the Orient and the south, impatient of rule and convention, free at last to utilize without constraint the exotic setting of a one time Spanish colony, is recorded here as in a portrait. . . . It is fortunate indeed that this great dream has been preserved in permanent form.

One could have applied "this great dream" to California itself, in its ebullient rise to stardom in the early and mid-twentieth century. Goodhue had drawn up the blueprints for how it wanted to see itself: at once arcadian and urbane, born of a luxuriant, romantic past, but simultaneously at the forefront of modernity, destined for unrivaled commercial, scientific, and military glory. It is fitting that Hale hired Goodhue in 1917 to design the new, 22-acre campus of the California Institute of Technology in Pasadena, of which he was director. Caltech became the world's leading institution in astronomy and rocket science, later spinning off the nearby Jet Propulsion Laboratory, where America's space program would be born and from where its interplanetary explorations are still designed, built, and directed. Goodhue created a Mediterranean-style baroque home for these most modern of scientists, as if their efforts were as old as civilization itself: a series of classically proportioned buildings with red-tiled roofs, upper loggias, and churrigueresque entrance facades, linked together by square-columned arcades like those of the local missions, and enclosing large courtyards planted with olive trees. Fronting the courtyards, Goodhue's trademark wit is on display: the arcades' column capitals are all sculptures, in cast concrete from originals done by Lee Lawrie, of whimsical figures demonstrating their trade, just as in the carvings on a medieval cathedral—a chemist with his beaker, a geographer measuring a globe, a scientist with a microscope, an engineer with an airplane engine (Southern California was a pioneering epicenter of the aircraft industry, and would build 70 percent of the US airframes flown in World War II). The scientists are paired with musicians: one playing a violin, another an accordion, another a horn; all with joyful and mischievous expressions.

What was new in Southern California in the first quarter of the twentieth century wasn't the romantic architecture of El Fureidis or Balboa Park—the entire nineteenth century and much of the eighteenth as well had been a frenzied bout of making new buildings in old styles. Nor was it the laying out of suburban neighborhoods with picturesque winding streets and landscaping to suggest Arcadia—these had long sprung up wherever commuter railroads allowed the well-to-do to escape the city for greener pastures. What was new at the level of the city and the region was a kind of city built on the illusion that it wasn't a city—the city dressed as the country, made up of detached houses, each scrupulously acting as if it had no neighbors, each a castle standing alone in pastoral splendor. There would be no verticality, except for church spires, Gothic or otherwise, and perhaps the odd castellated tower on a suburban villa. Even manufacturing would be housed in historically themed architecture, like the Samson Tire and Rubber plant in Los Angeles, then the largest single manufacturing building east of the Mississippi, built in 1929 in the alarming shape of a seventh-century BCE Assyrian castle complete with concrete griffins perched atop its crenellated battlements. Instead of the suburb fleeing to the edges of the city, the suburban ideal could colonize the city and, in some cases, replace it entirely.

When Goodhue perfected his dream-city model, he on the one hand demonstrated the architectural means for building entire cities as themed, romantic fantasy realms. More important, he and other architects of his persuasion ironically freed the romantic mode from its historical shackles by stripping away the last residues of the political project of the nineteenth-century urban reformers. They replaced the reformers' criticism of industrial capitalism with an unfettered narrative of personal romance, sustained by a cinematic level of illusion. Romantic architectural style, done right, could produce entertainment. People could imagine themselves as Spanish rancheros or Arthurian knights, not only in the privacy of their home, like Mr. Wemmick, but also while working, shopping, or socializing. The effect would be henceforth seamless. In building these environments, Goodhue and his collaborators and imitators achieved what the Arts and Crafts movement had failed to: they designed objects that truly had the power to transport us, at least in our imaginations, to a place far from the vexing problems of the modern city.

Leaving the city behind has its costs. Some are borne by those doing the leaving: isolation and disassociation from the broader community and its social, economic, and political life. More are borne by the community left behind, which subsidizes the ever-greater investment in roads, infrastructure, education, and resources that the suburb-city requires, while enjoying few of its benefits. The reformers' dreams of a better world came true for those in the middle classes who could afford the fare to travel to it, but left those lower down permanently stranded.

. . . .

In the next several years, Goodhue continued with the self-contained miniature-city mode, designing incongruously romantic Spanish-style base complexes near San Diego for the marine corps and the navy. He also went on to build a number of residences in the Santa Barbara area, including the Dater house (1915–18), a house for himself, and the Montecito Country Club (1916–17). He was not by any means alone in working in the Spanish style, but his pioneering work provided the prototypes for an emerging regional movement, dubbed the Spanish Colonial Revival, which gradually replaced the Arts and Crafts–derived Craftsman style so popular in California. Myron Hunt, originally from Massachusetts, worked with Elmer Grey as supervising architects for Goodhue at El Fureidis before the two partnered on the 1911 Beverly Hills Hotel, which became iconic; Hunt would go on to be one of Southern California's leading high-society architects, compiling 500 eclectic commissions in his career.

Goodhue's model of a full-immersion historical-architectural environment was taken to its extreme point in a single residence on the Central California coast at San Simeon, also known as Hearst Castle, the fantasy medieval hill town built by the architect Julia Morgan for the publishing magnate William Randolph Hearst. Beginning in 1919 and continuing for two decades, Morgan, who began her career in the Bay Area designing in the Arts and Crafts mode, designed Mediterranean Revival towers, chapels, villas, stables, swimming pools, and gardens, with the construction work done by a community of skilled craftsmen housed on the property in their own custom-built village. It was a perverse incarnation of the Arts and Crafts movement's workshop ideal, with everyone in the pay of one fabulously rich man, building a stage set for his hedonistic lifestyle of parties attended by a constant stream of Hollywood stars, European royalty, and business tycoons. San Simeon even resembles Goodhue's drawing of Xanadu, with a brilliant tiled pleasure dome at its center, though Morgan's version has two smaller bell towers to Goodhue's one massive spire.

The dream city of Balboa Park was what everyone wanted (not just Hearst), and it helped spawn entire cities where the Spanish mode dominated, especially those communities catering to the wealthy: Pasadena, Carmel-by-the-Sea, Holmby Hills and the Hollywood Hills in Los Angeles, La Jolla, Ojai, Palos Verdes, and Rancho Santa Fe, to begin the list. In Santa Barbara, it virtually became the law: after an earthquake in June 1925 leveled much of the downtown and its Victorian eclectic buildings, a group of preservationists and architects mobilized to establish a board of architectural review with permitting powers, which would all but compel builders to adopt the Spanish style. The signature development was the rehabilitation of the historic De La Guerra adobe with a landscaped public plaza fronted by a new Spanish-style building for the city newspaper and a pedestrian shopping mall called El Paseo (its design supervised by Carleton Winslow, Sr., who had also worked for Goodhue in San Diego), all of it melding a re-created Andalusian past with modern American resort-town commerce. El Paseo became an instant tourist attraction. Developers took note. Also in 1925, down the coast where Orange and San Diego counties meet south of Los Angeles, the developer of San Clemente made it officially a "Spanish Village by the Sea," building a mall, beach club, pool, golf course, playgrounds, school, and housing in the style, and placing a board-of-architectural-review requirement like Santa Barbara's in property deeds.

It worked: California boomed, especially the southern part of the state, where housing development quickly became the leading industry. A million and a half people moved to greater Los Angeles in the decade 1920–30 alone, and most of them came, at least in part, because they believed the promises that here every man and woman could have a castle of their own. Outside the most privileged, curated enclaves like Santa Barbara and Carmel, the romantic architectural palette was more flexible, offering a range of flavors, many of them unapologetically inspired by the movie business, whose sets were plainly visible. In Los Angeles, sets loomed over busy streets: the four-block Babylon from D. W. Griffith's *Intolerance* at the corner of Hollywood and Sunset boulevards; nearby, *Ben-Hur*'s Rome and *Robin Hood*'s Nottingham Castle occupied prime real estate. Builders echoed them, and the resulting architectural riot came to define the region—happily, to itself, and appallingly, to keepers of good taste from the East. No one captured the scorn better than the critic Edmund Wilson, who wrote in 1932, in a remarkably inventive bit of snobbish ridicule:

> The residential people of Los Angeles, are cultivated enervated people, lovers of mixturesque beauty—and they like to express their emotivation in homes that symphonize their favorite historical films, their best-loved movie actresses, their luckiest numerological combinations or their previous incarnations in old Greece, romantic Egypt, quaint Sussex or among the priestesses of love of old India. Here you will see a Pekinese pagoda made of fresh and crackly peanut brittle—there a snow-white marshmallow igloo—there a toothsome pink nougat in the Florentine manner, rich and delicious with embedded nuts. Yonder rears a clean pocket-size replica of heraldic Warwick Castle—yonder drowses a nausey old nance. A wee wonderful Swiss shilly-shally

> snuggles up beneath a bountiful bougainvillea which is by no means artificially colored. And there a hot little hacienda, a regular enchilada con queso with a roof made of rich red tomato sauce, barely lifts her long-lashed lavender shades on the soul of old Spanish days.

The Los Angeles regional growth machine quickly found its true enabler in the car: by 1930, the region had 800,000 cars (two for every three people); then it had millions. Today, Los Angeles County has 10 million people, part of a five-county supercity of more than 20 million, transported by nearly that many cars. It is the largest urbanized area in the United States, known to specialists as a "conurbation"—a city at regional scale, but an "urb" made up mostly of suburbia. Not all of it is romantic suburbia, by any means: there are endless gradations in types of buildings, streets, and people. The region-city was built over decades by tens or hundreds of thousands of actors, but most of it was built on the shared premise that the new city should not resemble the old city, but should instead be a kind of anticity. Planners followed strict rules limiting building heights and neighborhood density in a concerted effort to maintain the look and feel of the garden city ideal. Not being able to stanch the inflow of people, they succeeded instead in spreading their garden city over close to 10,000 urbanized square miles from the Mexican border north to Santa Barbara—a distance of 200 miles—a dream megalopolis that Bertram Goodhue would not easily have recognized as his inspired offspring.

1. Castles

Field Guide: Romantic Suburbia

Diagnostics:

- "Picturesque" historical architecture, heavy on ornament and detail; modeled on preindustrial, often rural examples such as castles, manors, and villas; also churches.
- Mostly residential, with prominent shopping centers; industry and mixed-use discouraged.
- Preferably detached houses, emphasizing separation from neighbors. Where houses touch, the illusion of separation is attempted.
- Curving streets and naturalistic landscaping. Too dispersed for walking. Car-dependent.
- Controlled access, through boundaries or greenbelts; often gated.

Examples:

- Spanish Colonial Revival districts of Santa Barbara, California; widespread in Carmel, Los Angeles, San Clemente, and elsewhere; Palm Beach, Boca Raton, and other parts of Florida.
- Nineteenth- and early twentieth-century railroad/trolley suburbs, typical around Eastern and Midwestern cities: Brookline, Massachusetts; Forest Hills Gardens, Queens, New York; Shaker Heights, Ohio; Riverside, Illinois.
- Post-1920s automobile suburbs with mixed architectural styles: Westchester County, New York; Beverly Hills, California.
- London: Hampstead garden suburb, Bedford Park, Blackheath.

Variants:

- Mixed forms: combining eclectic, picturesque historical architecture with more modern kinds; combining pockets of curving streets with gridded or semigridded town plans, and mid- to high-rise cities.
- Watered-down forms: tract developments with crude theming, shared historical theming, often gated, with curving streets and landscaping, but with denser, smaller lots, houses closer together; architecture more pastiche, lower quality of materials and detail: Rancho Cucamonga, California; Orange County; Beijing, China.

Images:

Rockefeller Chapel, University of Chicago (1925–28). Architect: Bertram Goodhue. *Mather G. Bisanz*

Los Angeles Public Library (ca. 1935). Architect: Bertram Goodhue.

El Fureidis, Montecito, California (1906). Architect: Bertram Goodhue. *Tai Kerbs*

California State Building, Panama-California Exposition, San Diego, California (1915). Architect: Bertram Goodhue. An example of the "churrigueresque" baroque Spanish style.

Courtyard, House of Hospitality, Panama-California Exposition, San Diego, California (1915). Architect: Bertram Goodhue.

Santa Barbara County courthouse (1929), Santa Barbara, California. Architect: William Mooser.

Chapter [2]

Monuments

Daniel Burnham and the Ordered City

The happier people of the rising City Beautiful will grow in love for it, in pride in it. They will be better citizens, because better instructed, more artistic, and filled with civic pride.

—Charles Mulford Robinson

At his death in 1912, there was no more famous architect in America than Daniel Hudson Burnham. Acclaimed as a father of the skyscraper, planner of the Chicago World's Fair of 1893 and the revamped National Mall in Washington, DC, he was sought after, wealthy, well connected, and toasted all over the United States and beyond. President Taft called him "one of the foremost architects of the world." His oft-repeated credo, "Make no little plans, they have no magic to stir men's blood. . . . Make big plans . . . remembering that a noble, logical diagram once recorded will never die, but long after we are gone will be a living thing asserting itself with ever growing consistency," perfectly embodied Americans' aspirations and burgeoning sense of national pride in the era of Manifest Destiny. A formalist and neoclassicist carrying nineteenth-century styles and ideas into the twentieth, Burnham had his detractors, notably those mavericks among his colleagues who subscribed to the modernizing tendencies that would soon after his death eclipse his fame and relegate his reputation to the dustbin of architectural history. Yet even the foremost modernist of his time, Frank Lloyd Wright, allowed that Burnham "was not a creative architect, but he was a great man."

His influence was matched by his size: over six feet tall and always impeccably dressed, he prided himself on physical fitness. A fencer in his youth, he played golf and kept a gym in his office later in life. He liked food and drink and fine accommodations in quantity—he bought his cigars in bulk—and as success and celebrity mounted his girth grew in step with his importance and reputed pomposity. Burnham built big buildings—including some of the first skyscrapers, such as the 10-story Montauk Building of 1882 (probably the first building ever to be called one), or the 16-story Monadnock Building of 1889–91. They were perfect expressions and containers for the swelling commercial power of Chicago, the self-described City of Big Shoulders, and as he moved on to put up big buildings in New York, San Francisco, and elsewhere, the skyscraper became the architectural symbol of America's imperial course at the turn of what would come to be called the American Century. On the heels of his triumph as director of the Chicago fair, officially the World's Columbian Exposition of 1893, Burnham became America's preeminent urban planner, redrawing the circulatory and symbolic maps of cities from Chicago to San Francisco to newly conquered Manila with an unforeseen grandeur, ostentatiously claiming the legacy of classical Greece and Rome and Renaissance Europe as rising America's inheritance. His skyscrapers inspired architects all over the world, most ironically the triumphant modernists who would otherwise disparage him. His reinvention of imperial city planning became the model for modern city builders of all political stripes: fascists, communists, and democrats; despots, oppressors, and progressive improvers alike. Daniel Burnham gave the twentieth century two of its universal urban forms.

Burnham was, in his beginnings, an ordinary American of his era. He was born in 1846 near Henderson, New York, into a family of colonial English stock, dating from the 1635 arrival of Thomas Burnham from Norwich, England, in Ipswich, Massachusetts. After unsuccessful stints as a merchant in New York State, his father moved the family to Chicago when Daniel was nine, in 1855, and entered the wholesale drug business. There, his fortunes rose, enough that in 10 years he became president of the powerful Chicago Mercantile Association. Daniel attended Chicago's Central High School, where he was remembered as athletic and handsome, and preferring drawing to schoolwork. A contemporary recalled that he "rarely studied and was always censured for his negligence. He was a tall . . . fellow, much too large for his age." His parents paid for drawing lessons and sent him east for two years to prepare for college entrance exams. He failed both Harvard's and Yale's. For four months he worked as a salesman, then, steered by tutors and older acquaintances, and his own love of drawing, he turned to architecture, gaining a place as an apprentice at the Loring and Jenney firm in Chicago. Clearly, he was inspired by the work, seeing in it a path to personal glory: "I shall try to become the greatest architect in the city or country," he wrote. "Nothing less will be near the mark I have set for myself, and I am not afraid but that I can become so. There needs but one thing. A determined and persistent effort."

But at that time Burnham wasn't ready to give it. Leaving Chicago, he tried his luck prospecting in a Nevada silver camp, staking a claim but finding no riches. Then he tried politics, encouraged by an older mentor, running for Nevada state representative from White Pine County. Unsuccessful, at the age of 24 he returned home, admitting his unsettled nature: "There is a family tendency to get tired of doing the same thing very long." Again he worked in several architecture firms, before being hired by Carter, Drake and Wight, on the recommendation to partner Peter Wight from Burnham's father, who, Wight later recalled, "was very desirous that Dan should be cured of his roving disposition."

In Wight Burnham found the mentor he sought, steadying him for that "determined and persistent effort" he had vowed to make, and in Wight's office he met the partner he needed: John Root. A young man from Georgia who had spent the Civil War in Liverpool, studying music with England's top organist and passing the Oxford entrance exams before being summoned home by his parents at the war's end, Root studied engineering at New York University, then worked in the architectural offices of James Renwick (architect of Saint Patrick's Cathedral) and J. B. Snook (architect of Grand Central Terminal). In 1872, he came to Chicago, becoming head draftsman in Wight's firm—part of a wave of people moving to the city to take part in the rebuilding after the Great Fire of 1871. Four years younger than Burnham, John Root was in many ways his complement. Where Root was highly cultured and trained but unambitious, Burnham was the opposite. Sensing reciprocal qualities, the two opened their own practice in 1873 with a handful of jobs, before the nationwide economic Panic of '73 constricted the building market. They scraped by on part-time work, including Root playing organ at the First Presbyterian Church.

Then began Burnham's rise in society, and with it, the firm's. Introduced by a contemporary to the stockyards magnate John B. Sherman, Burnham landed the firm's first big commission, a three-and-a-half-story mansion on fashionable Prairie Avenue, with all the Victorian gewgaws: steeply pitched roofs, tall dormers, fluted brick chimneys, and capitals; the building was completed in 1874. Other commissions followed, with an increasingly prominent roster of clients, as did invitations to social events and clubs. In 1876, he married Sherman's daughter Margaret. Between the architecture's opulence and Burnham's likability and charisma, the fit with high society was perfect. Within the firm, the division of labor saw Root handling the details on paper and Burnham the clients. Burnham's ease with people was near legendary. Frank Lloyd Wright said that "his powerful personality was supreme." Wright's mentor Louis Sullivan, another celebrated Chicago architect who, with his partner Dankmar Adler, would also be central to the invention of the skyscraper, and was something of a rival of Burnham's, recalled his first impression of him as "a dreamer, a man of fixed determination and strong will . . . of large wholesome, effective presence, a shade pompous . . . a man who readily opened his heart if one were sympathetic." In the judgments of critics and historians, Burnham has often been compared unfavorably with Sullivan, with the latter considered a proto-modernist master, and the former a glorified businessman, relying on his partner Root to supply the buildings. But it is clear that Burnham, with his knowledge of their clients' needs, played the major part in defining each project's initial program and floor plans, and actively interacted with Root bringing the drawings to completion.

In the 1870s and '80s, Burnham and Root's buildings drew from the uproarious mixture of styles that marked Victorian architecture: Romanesque, Queen Anne, and French Renaissance among them.

The era's eclecticism made no pretences to saving the world, as Root acknowledged when he joked that the Victorian mode would better be called the "cathartic," Romanesque the "dropsical," and Queen Anne the "tubercular." Combining and varying the styles as most contemporary firms did, Burnham and Root built big, stolid houses, churches of several denominations, clubhouses, hotels, and railroad stations in Chicago and all over the Midwest. Notable among them were the original Chicago Art Institute building on Michigan Avenue and the Montezuma Hotel in far-off Las Vegas, New Mexico, a fanciful pink confection with an incongruous onion dome, paid for by the Atchison, Topeka, and Santa Fe Railroad, which over the years hosted the outlaw Jesse James, Japanese emperor Hirohito, and US presidents Ulysses Grant, Rutherford Hayes, and Theodore Roosevelt.

In the 1880s, Burnham and Root were perfectly placed to be the leading architects of Chicago's extraordinary commercial rise. The city was an avatar of the urban transformation of the United States between the end of the Civil War and the turn of the century, as population growth combined with technological progress on the farm, in manufacturing, and crucially, in transportation, to drive the growth of cities. In 1870, the country's population was 40 million, 60 percent of them farmers; by 1900, the population had reached 70 million, with just 37 percent of them farmers. An explosion of railroads, financed by Wall Street in a roller coaster of booms and busts, transformed agriculture, allowing bigger farms in the West to industrially grow commodities for distribution to the cities, instead of relying on smaller, diversified, local markets. Rail miles grew from 30,000 in 1870 to 170,000 in 1900. Chicago benefited the most from this incredible flowering of iron, making itself the center of a vast web of tracks across 10 or more states that brought lumber from Northern forests and wheat and corn from the prairies into the city, then shipped them out again, along with meat, from animals grain-fed in Chicago stockyards and processed in Chicago slaughterhouses, shipped to the East in refrigerated railcars invented by a Chicago corporation. The city's growth was astonishing, doubling from half a million in 1880 to over 1 million a decade later.

For architects, Chicago provided a perfect confluence of extraordinary conditions for innovation. As business and population grew, land rents spiraled: the cost of a quarter acre in the city center jumped from $130,000 in 1880 to $900,000 in 1890, and $1 million in 1891. Unconstrained by legal height restrictions that existed in other cities, especially in Europe, architects in Chicago built ever higher for their corporate clients on ever-smaller parcels, taking advantage of a torrent of inventions: elevators, first drawn by horses, from 1853 by steam-driven hydraulics, and from 1889 by electric motors; electric lightbulbs; and telephones. Taller buildings concentrated and multiplied economic activity, driving land values even higher. Burnham quickly revealed a talent for efficiently managing large projects and his own growing firm. To Sullivan, he embodied the spirit of the age: "During this period there was well underway the formation of mergers, combinations and trusts in the industrial world. The only architect in Chicago to catch the significance of this movement was Daniel Burnham, for in its tendency towards bigness, organization, delegation and intense commercialism, he sensed the reciprocal workings of his own mind."

In 1881, the firm completed the 7-story Grannis Building, and in 1882, the 10-story Montauk; encouraged by its practical and frugal owner, Burnham and Root constructed the building with plain brick, minimal decoration, and a flat roof. Its austere style indicated a new and pointedly modern direction, dictated not by fashion but by commercial logic. These buildings tested the limits of tall masonry structures, as their walls grew ever thicker at the base to support their weight. Gradually, advances in affordable steel beams, pioneered by the railroad industry, were introduced into Chicago's rising downtown, making walls thinner, and, when they were encased in concrete, buildings fire resistant. The Home Insurance Building of 1885 by Burnham's old employer William Jenney showed the way toward a near-fully steel-framed structure, and Burnham and Root's 10-story Rand-McNally Building (1888–90) is often considered the first to be entirely steel supported. The firm followed it with more buildings in Chicago, San Francisco, and Atlanta. As the buildings grew taller and more technologically complex, their surface decoration grew simpler; the 16-story Monadnock Building in Chicago (1889–92), with its bay windows and clean facade, was so spare that many compared it to a machine—prefiguring twentieth-century architecture's favorite metaphor. Sullivan, who more than any Chicago architect is generally thought to have anticipated modernism, found it stunning: "An amazing cliff of brickwork, rising sheer and stark, with a subtlety of line and surface, a direct singleness of purpose, that gave one the thrill of romance."

If Burnham's skyscrapers with John Root boldly pointed toward the possibilities of the future, his other contribution to the modern city, what came to be called City Beautiful planning, was backward looking: to Europe, to the classical past, to a conservatism that was at least in part a product of American cultural insecurities.

After the triumphant 1889 Paris World's Fair, the fifteenth and biggest of the world's fairs held in the second half of the nineteenth century, the United States vowed to be next, proposing a fair for 1893 to mark the 400th anniversary of Columbus's discovery of America.

After much lobbying, upstart Chicago topped competing bids from New York and Saint Louis, and in 1890 the countdown began. The sponsoring group hired Frederick Law Olmsted, the most famous landscape architect in America, and his partner Henry Codman to choose a site, and sent Daniel Burnham along as local adviser. They chose an undeveloped lakefront parcel on the south side of the city called Jackson Park. Olmsted and Codman drew an initial plan, with Burnham and Root made consulting architects. It was to be a stupendous undertaking: a real city, on more than 600 acres—four times bigger than the Paris fair—complete with sewers, gas, electricity, water, lighting, streets, and drainage, plus police, a fire department, medical services, insurance, communications, and more. In October, Burnham was named chief of construction (later called director of works), in charge of all infrastructure, design, engineering, and construction, all to be finished in three years.

For the architecture, the initial idea was to copy the giant glass-and-iron buildings used since London's Crystal Palace of 1851. Instead, Burnham wanted to "abandon the conservatory aspect of older exhibitions and . . . suggest permanent buildings—a dream city," he wrote. Each major portion of the exhibition would have its own building, and the decision was made to bring in different architects for each. Burnham settled on five of the East's most prominent and talked them into accepting. Three were New Yorkers: Richard Hunt would design the central Administration Building; Charles McKim of McKim, Mead and White the Agriculture Building; George Post the Manufactures and Liberal Arts Building. Peabody and Stearns of Boston would design Machinery Hall, and Van Brunt and Howe of Kansas City, Electricity Hall. Burnham and Root would be executive architects, handling the actual construction duties. Under pressure from the Chicago organizers, five additional local architects were chosen for other buildings, including Adler and Sullivan for the Transportation Building.

In 1891, John Root died of pneumonia, increasing the already unimaginable pressure on Burnham. All the architects were called to Chicago for a week, along with Olmsted, and the sculptor Augustus Saint-Gaudens, added as the arts consultant. There, a decision was made to design the entire fair in a single style—not the colorful Victorian eclectic of earlier fairs, nor the new Chicago style, so modern and machine inspired, and arguably best suited for the machinery and inventions that would be displayed. Instead, all the buildings would be in the neoclassical style, with a single cornice line and common details. Since being supplanted earlier in the century by Gothic and other styles, neoclassical was newly fashionable: many of the fair's architects had trained at the École des Beaux-Arts in Paris, where it was de rigueur, and the New York and New England cultural elite were in the throes of fashioning an "American Renaissance," drawing on European neoclassicism in several arts. The style's reappearance was a manifestation both of Americans' sense of their rising role in the world and of the cultural anxiety it provoked. More and more Americans, made wealthy by the industrial juggernaut that fed the Gilded Age—wealth grew by $20 billion in the decade between 1883 and 1893—took grand tours in Europe, where the Belle Epoque's grandeur and refinement stunned them, provoking both ambition for their own country and embarrassment by it on their return. In 1904, Henry James, the American writer self-exiled in Great Britain from the age of 26 until his death at 73 in 1916, published *The American Scene*, his classic excoriation of the crassness and misbegotten aesthetics of the newly rich after returning to his native land. Chicago, grown so recently and so big, with its stockyards, rail yards, packing houses, and filthy working-class districts all jammed together, evoked particular horror in those recently returned from abroad. Such was the response of the young, artistic hero Truesdale Marshall in Henry Blake Fuller's 1895 novel, *With the Procession*, on his return from a European tour: Chicago was a "hideous monster, a piteous, floundering monster too. It almost called for tears. Nowhere a more tireless activity, yet nowhere a result so pitifully grotesque, gruesome, appalling."

Daniel Burnham's Chicago fair would be otherwise. The design took shape: the main buildings would be arranged around a large basin of water called the Court of Honor, headed by a 600-foot-long peristyle, or Roman columned court, with each column standing for a US state; all would be painted white and festooned with sculptures and ironwork. A pier would extend into Lake Michigan for arrival by boat. A lagoon was to be plied by colorful ships from all over the world. Expectations were high, not least among the architects, sculptors, and painters assembled for the job. (Saint-Gaudens expressed their self-consciousness at a February 1891 meeting of the team: "Look here," he said. "Do you realize that this is the greatest meeting of artists since the 15th century?") Then the work began. In the spring, Burnham moved into a construction headquarters in tents erected on-site and took command of a force of thousands of workers laboring year-round, confronting setbacks, such as snowstorms collapsing roofs and the marshy ground challenging foundations. Using steel frames covered in wood and plaster and painted to resemble stone, the fairgrounds rose on the Lake Michigan shore like a giant stage set, thrown up cheaply and in under two years.

The World's Columbian Exposition opened May 1, 1893, with President Grover Cleveland in attendance. The scene itself was the star attraction: named the White City in the press because of the buildings' bright white paint and the electric bulbs that blazed throughout the grounds at night—the largest electrical system ever mounted, driven by George Westinghouse's enormous dynamos in the Hall of Electricity. Including the main buildings and the pavilions of US states and territories and 46 foreign countries, 200 buildings showed off the agricultural, cultural, and technological wonders of the world: one could see Krupp's artillery in the German pavilion, Nikola Tesla's neon and phosphorescent displays, Eadweard Muybridge's moving pictures, George Ferris's first Ferris wheel, a belly dancer in a "street in Cairo," and cowboys and Indians re-creating their recent wars in Buffalo Bill's *Wild West Show*. In the fair's six-month run, 21,480,141 people came—equivalent to roughly half the population of the United States.

The fair was a coming-out party not just for Chicago, but for America. The memoirist Henry Adams wrote that "Chicago was the first expression of American thought as a unity; one must start there." Inside, the historian Frederick Jackson Turner had given his famous lecture announcing the end of the American frontier, on the basis of data from the census of 1890, signaling that a new epoch in the nation's life was beginning. The country was rapidly becoming urban and industrial, even in its agriculture, and increasingly divided: by race, ethnicity, and religion, owing to surging immigration; and by wealth, since income disparities grew to unprecedented levels as robber baron capitalism concentrated the nation's riches in proportionally fewer and fewer hands. The achievement of the fair itself was certainly an expression of what would become known as Progressivism: a wide and multipronged reform movement that had two major vectors. The first was toward efficiency: in management and government, targeting corruption, and encouraging civic participation, scientific methods, better health and safety, hygiene, and infrastructure, and the conservation of natural resources by establishing national parks, wildlife reserves, and forests. The second vector was toward social justice: scaling back child labor, advancing women's suffrage, and ameliorating the condition of the poor. Much thought and effort went into reforming urban slums, which by the late nineteenth century had become an appalling spectacle for members of the middle and upper classes, as they had earlier in Great Britain. Reporters and reformers who ventured into city slums, densely populated by mostly European immigrants, returned with stories and images of people living heaped up in dark, filthy rooms, too often with little or no sanitation, heat, clean water, or clean air. Fascination and fear, of otherness and of class revolution, combined with the conviction that the slums constituted a mortal threat to American civilization, prompted architectural attempts at reform. Jacob Riis's photographs of New York's Lower East Side, published in 1889–90 as *How the Other Half Lives*, led to experiments in building "model" tenements, but these meager design changes did nothing to alter the economic conditions under which the poor struggled.

The architecture of the fair was no less improvement minded, but it wasn't the expression of interest in social justice or better housing for the poor. It was part of a completely different tradition: of designing the civic realm in order to inspire, educate, and instill proper values by example. A beautiful environment inculcates civic virtue, the theory goes. In this tradition neoclassical architecture had long been the preferred way. Many saw in the fair the promise of a powerful, even mystical, influence on the hoi polloi: the journalist Henry Demarest Lloyd thought it "revealed to the people, possibilities of social beauty, utility, and harmony of which they had not been able even to dream. No such vision could otherwise have entered into the prosaic drudgery of their lives, and it will be felt in their development in the third or fourth generation." It was not without controversy:

some contemporary critics called it backward looking; Louis Sullivan later griped that the choice of neoclassical had set American architecture back by a generation, eschewing "the architecture of Democracy" (by which he clearly meant to describe his own work) for a "feudal," "imperial," and "Philistine" style.

But for most, the Chicago fair was a revelation. Burnham himself wrote, "People are no longer ignorant regarding architectural matters. They have been awakened through the display of the world's Columbian Exposition of 1893." He reaped much of the very public accolades, including receiving honorary degrees from both Yale and Harvard in 1894, and Northwestern's first honorary PhD. The American Institute of Architects made him president in 1893 and 1894. In 1896, he and his wife, Margaret, traveled to Europe and the Mediterranean, visiting France, Italy, Malta, Carthage, Egypt, and Athens. There, he imbibed the classical stream at its sources and resolved to bring it home. At the Acropolis in Athens he had a reverie: "It was a perfect evening. . . . [My wife and I] sat entranced, speechless on the rocks, amid fallen columns. . . . I have the spirit of Greece once and forever stamped on my soul. It is the blue flower, the rest of life must be the dream and this land of Greece the reality."

In reality, Burnham missed the message in the broken stones of a long-fallen civilization. If the architecture of the fair was in a stylistic sense continuous with that of the young American republic, in another its implications were very different: the fair's grandiose pseudo-Roman pomp was an expression of the new, muscular capitalism organizing and dominating the economy. The United States was at a combustible point: tensions ran deep in post–Civil War society, between races, classes, native-born and immigrants, Protestants and Catholics, capitalists and unions. The country's growth had been built on violence: of slavery, of expansion through war against European powers, Indian tribes, and Mexico. An imperial swagger permeated an America besotted with the ideology of Manifest Destiny, and soon the United States would become an imperial power, seizing the Hawaiian kingdom and then Spain's last Caribbean colonies and the far-off Philippines. Against the backdrop of the Chicago Haymarket riot in 1886, the Homestead strike in 1892, and the Pullman strike in 1894, tension had been palpable at the fair, as unions tried to organize the workforce and Burnham determinedly purged the ranks. After it closed, much of the fair burned to the ground under suspicious circumstances. But this did little to tarnish the White City's—and Burnham's—triumph. In the national election of 1896, the conservative, pro-business Republican William McKinley defeated the populist Democrat William Jennings Bryan, signaling the end of the old, more liberal Republican Party of Abraham Lincoln and the dominance of urban, corporate interests over traditional agrarian ones. The historian T. J. Jackson Lears summed up: "What had triumphed in 1896 was not the White City, but the machines in it."

Post-fair, Burnham's practice, renamed D. H. Burnham and Company after Root's death, boomed, with an office on the Loop in Chicago and a second in New York opened in 1900, while he and his wife, Margaret, lived the suburban life in Evanston, Illinois. It was no longer a typical architecture atelier, but grew into a large firm, mirroring its corporate clients. The firm built in every mode: neoclassical banks and libraries, eclectic railroad stations, opulent department stores, and more skyscrapers, including the 1894 Reliant Building in Chicago and the 1903 Flatiron Building in New York, then the tallest in the world. All were big, grand, and assured, like their architect and the clientele he served. At Burnham's death in 1912, D. H. Burnham and Company employed 180 men and had realized 200 buildings. It was an extraordinary record of success.

In 1901, Burnham was approached with his next major challenge: redesigning the heart of the nation's capital. Laid out on paper by the French engineer Pierre L'Enfant in 1791 as a grand Renaissance diagram of regular blocks striated with tree-lined avenues and diagonals radiating from a series of statue-studded squares, the whole focused on the sites of the White House and the Capitol Building. Down the center of the plan, extending west from Capitol Hill, was a ceremonial National Mall, stretching two miles to the Potomac River. The city, famously, had been little more than swamp and pasture at its birth, when George Washington was president and Thomas Jefferson secretary of state, and it had filled in only slowly over the years, often with uses not conforming to L'Enfant's vision. The Mall had never been finished. A portion had been laid out as a meandering, picturesque "English Garden" by Alexander Jackson Downing in 1850, but had mostly succumbed to neglect in the half century that had passed. In the Civil War, troops had been quartered on it, then parts used for pasture, a lumberyard, and a railroad yard and tracks, still in 1901—a perfect symbol of America's rapid, often sloppy industrialization, and the acute need to rethink and rebuild public spaces.

In 1901, Michigan senator James McMillan, chairman of the Senate's District of Columbia committee, organized a three-man planning commission to propose solutions for the Mall, asking Burnham to chair it, with Frederick Law Olmsted, Jr. (the master's son), and Charles McKim, who then requested that Augustus Saint-Gaudens be added. They would work pro bono. They faced at least one practical problem: soggy ground at the projected location of the Washington Monument, at the point where the axes of the White House and the Capitol crossed, made it impossible to lay the foundation. There were also aesthetic questions, and deciding the extent of the intervention they contemplated. Burnham, true to his credo, wanted to make big plans, for he thought the city would someday attract a more prosperous population than it then had. He wrote to Olmsted: "My own belief is that instead of arranging for less, we should plan for rather more extensive treatment than we are likely to find in any other city. Washington is likely to grow very rapidly from this time on, and be the home of all the wealthy people of the United States." He proposed to Senator McMillan that the government pay for a research trip to Europe: "How else can we refresh our minds, except by seeing, with the Washington work in view, all those things done by others in the same line?" Beforehand, the team toured the colonial Virginia estates of Stratford, Carter's Grove, and Berkeley, and the town of Williamsburg, which had a plan like Washington, DC, in miniature and may have been a model for L'Enfant. Then, they toured Paris, Rome, Venice, Vienna, Budapest, Paris again, and the royal gardens at Versailles, Fontainebleau, and Vaux-le-Vicomte. Burnham went alone to Frankfurt and Berlin, before rejoining the others in London, and seeing Hyde Park and Hampton Court, then Eton and Oxford.

These places were mostly royal and imperial, and all were aristocratic—odd inspirations for the capital of a republic. But this discordance stems from the basic contradictions in the Founding Fathers' adoption of neoclassical style—contradictions simmering at the heart of the American experience. Burnham, with his upper-class identification, merely accentuated it. On one hand, it was perfectly consonant: the Renaissance's early interest in reviving classical forms was part of the Enlightenment project of reason and science seeking clarity and harmony in the laws of an ordered universe—the philosophical bedrock of republican thought. City plans of the era emphasized the aspect of modernization, opening up the chaotic, plague-ridden, and flammable medieval city and replacing it with one designed to better move people, goods, and troops when necessary. Such were the proposals for rebuilding London after the Great Fire of 1666 by Christopher Wren, Robert Hooke, and John Evelyn, with their regular squares and diagonal avenues. Their authors believed that their more rational plans would confer economic benefits and practical and social advantages: the new city was also a mark of refinement, of a piece with the general aspiration to display the signs of discernment, class, wealth, grace, and virtue.

On the other hand, classical style speaks loudly of the old order, the ancien régime, whether the nobility, royalty, or the Church. By definition, the Renaissance city is a revival of Rome. The question faced by everyone who has been tempted to emulate it is whether one is reviving the republican Rome of the Senate or the imperial Rome of the Caesars. Versailles is a prime example of this conflict: a place of intense aesthetic and horticultural prowess, of science, beauty, whimsy, and wit; but also, irreducibly, a stark diagram of the power of the French monarch and his state, a demonstration and reminder of its military, religious, economic, and administrative control over the territory claimed by the diagonals radiating outward from the palace at its center. The neoclassical city (for there is no truly classical city, as all the examples we have of past urban glories were already imitations of something that came before) is a show of force. Roman planning derived in good part from the forms of its military encampments. European Renaissance planning owed much to advances in fortifications and the building of garrison towns, bastides in France, and their counterparts in nearly every country—with their ramparts, plazas for drilling troops, and carefully planned sight lines and avenues for fields of fire and advancing forces. Machiavelli, theorist of the city and military strategist, and Sébastien Vauban, royal engineer of defenses and ways to undermine them, helped lay the foundations for the cities Daniel Burnham dreamed of building.

Baron Haussmann's massive reshaping of Paris for Emperor Napoleon III from 1853 to 1870, clearing congested older neighborhoods for wide boulevards, avenues, parks, bridges, and monumental civic buildings and railroad stations, served multiple objectives. The grand boulevards and monuments that punctuated them reflected glory on the state. Straight roads opened the districts of the poor to quick troop movements and clear fields of fire—certainly of value given the six uprisings in Paris in previous decades. Streets designed for faster vehicle traffic and a new central market, Les Halles, boosted commerce. Underground water pipes and sewers made the city healthier, while lavish parks, squares, and mile after mile of regular architecture and cornice lines gave it an aesthetic updating.

The entire process expedited the removal of the poor from the central city and refashioned it into a place that could, in Burnham's words, "be the home of all the wealthy people." One way gentrification proceeds is by remaking the city into a museum of its own past, a monument to itself, and indirectly, to those clever enough to live there. The past, re-created or conserved, confers legitimacy on the powers of the present. The nineteenth century in Western culture was an ongoing crisis of time: because of the continuous dislocations and traumas of modernity, culture was obsessed with finding the right memories, trying to make sense of, or take refuge from, the present by recovering a golden age, or at least finding a usable past for a guide. The great advances of nineteenth-century science were those of deep time: Charles Lyell's geology, Charles Darwin's evolution, Johann Winckelmann's archeology; its social sciences were of history and memory: Hegel and Marx, William James, Freud, and Henri Bergson. The modern museum was invented, too; it was no longer simply a place where collections of objects from the past were staged but a stage set in its own right. The Louvre in Paris, the British Museum in London, the Kunsthistorisches Museum in Vienna, and the Metropolitan Museum of Art in New York are just some examples of museum-monuments to rising imperial cities and their ruling classes set up in the last part of the nineteenth century.

In Washington, the commissioners decided on a big move: the McMillan Plan of 1901, which greatly expanded and rethought L'Enfant's scheme, creating a Versailles-like diamond diagram from the White House along Pennsylvania Avenue to the Capitol, defined by tree-lined boulevards, monuments, and parks. They solved the problem of the Washington Monument with the practical accommodation of shifting the main east-west axis of the Mall a few degrees to the south, where there was firmer ground. The center of the Mall would be a 300-foot-wide grass strip, flanked by avenues lined with four rows of elms. Behind these on either side were to be lines of stolid, neoclassical government buildings. Beyond the monument the commissioners added a reflecting pool, terminating at a massive memorial to President Lincoln, to be built on land reclaimed from the Potomac by dredging a large basin from the marshes (the Tidal Basin). The north-south axis would have its own terminus in another memorial, to Thomas Jefferson. Both presidents would be turned into larger-than-life monuments, each housed in his own colonnaded pagan temple. The problem of the railroad was solved nicely when Alexander Cassatt, the president of the Pennsylvania Railroad (and brother of the Impressionist painter Mary Cassatt), bought his competitor Baltimore and Ohio Railroad, and agreed to move his tracks off the Mall and into a new Union Station, to be designed by Burnham. The station was completed in 1907, five blocks from the Capitol. Burnham pulled out all the stops for his client, producing a magnificent neo-Roman building that is one of the architect's most impressive and one of the signature structures of the Gilded Age, its entry and 600-foot-long facade studded with sculptures by Saint-Gaudens, all taken from the Arch of Constantine, its 96-foot-high interior vaulting from the Baths of Diocletian. As for the McMillan Plan, Congress was resistant to its ambition and expense, and even with the support of President Roosevelt, years of politicking were required by Burnham and the senator, and finally President Taft's creation of the Fine Arts Commission in 1910 with Burnham as its chair, before the plan could be executed. Its last major element, the Lincoln Memorial, was finished in 1922.

The work of Daniel Burnham and his collaborators formed part of a broad movement advancing aesthetic improvement as a cure for the nation's urban ills. It extended from local garden and civic clubs seeking in the common parlance to "beautify" and "embellish" their neighborhoods with plantings, fountains, and civic art, all the way to expensive, official city-planning efforts to build boulevards, parks, and grand neoclassical buildings. Leadership was provided by Charles Mulford Robinson, a journalist from Rochester, New York, who had been so impressed by the Chicago fair that he made himself over into an apostle for what he called the City Beautiful, writing articles, lecturing, and publishing in 1901 *The Improvement of Cities and Towns*, which became the popular bible of the movement. His watchwords were "civic beauty" and "adornment," to be accomplished with public art, gardens, tree planting, street paving and lighting, by suppressing litter and advertising, and by zoning unsightly or polluting industries to other parts of town. He extolled the program as a way to simultaneously encourage a beautiful, healthy environment and "good citizenship" in its participants. A staunch Progressive, he advocated the appointment of expert committees like the McMillan Commission—and offered his own services. He was hired by local City Beautiful committees and cities themselves to prepare scores of reports and plans, including those for Detroit, Denver, Oakland, and Honolulu, just after its forcible annexation by the United States. His 1907 master plan for Los Angeles proposing to make it into the "Paris of America," by crisscrossing it with grand landscaped boulevards, was never acted on. His far more modest plan for Santa Barbara in 1909 was mostly adopted.

Robinson gave "the dream city" of the Chicago World's Fair credit for having "revealed a yearning toward a condition which we had not reached," in the United States, and "immensely strengthened, quickened, and encouraged . . . a desire that was arising out of the larger wealth, the commoner travel, and the provision of the essentials of life." To critics, the movement was, like its model, a sham, meant to distract from the real urban problems that existed behind the alluring neoclassical curtain. Lewis Mumford wrote that "the evil of the World's Fair triumph was that it . . . introduced the notion of the City Beautiful as a sort of municipal cosmetic, and reduced the work of the architect to that of putting a pleasing front upon the scrappy building, upon the monotonous streets and the mean houses, that characterized vast areas in the newer and larger cities."

Nevertheless, the movement's vision had huge appeal across the political spectrum. In Edward Bellamy's runaway best seller, *Looking Backward*, first published in 1888 and by 1900 the second-highest-selling American book after Harriet Beecher Stowe's *Uncle Tom's Cabin*, the protagonist Julian West, after falling asleep in 1887, wakes in the year 2001 to an achieved American Utopia of decidedly City Beautiful character: "At my feet lay a great city. Miles of broad streets, shaded by trees and lined with fine bldgs . . . stretched in every direction. Each quarter contained large open squares filled with trees, along which statues glistened and fountains flashed in the late-afternoon sun. Public buildings of a colossal size and architectural grandeur . . . raised their stately piles on every side." Bellamy's Utopia was explicitly socialist—proof that the appeal of grandeur, size, and stately monumentalism has no political boundaries.

City Beautiful proponents were likely to be motivated by one or more overlapping beliefs: a deep-seated faith in the power of things to change people, a rational-seeming faith in science and management to better order their affairs; a medicalized view of the city as a body subject to blights and cancers, which must be cured with treatments to out the diseased growth and encourage healthy development; plain wishful thinking; and varying degrees of self-interest. Robinson was clear on the last argument: "Municipal advance on aesthetic lines has been supported by an interesting economic argument. This was not needed, but of late it has been so much referred to that it cannot be properly passed over. It expresses the value of civic attractiveness in dollars." He went on: "The lesson that was plainly sought, and as plainly taught, is that it is financially worth while for a city to make itself attractive; lovely to look upon, comfortable to live in, inspiring and interesting."

Americans hardly needed convincing. The evidence remains visible in parks, squares, monuments, avenues, and grand neoclassical buildings: those thousands of white-columned banks, courthouses, and museums in towns and cities across the land.

Burnham himself remained much in demand. In 1902, Cleveland's Progressive mayor, Tom L. Johnson, hired him to chair a commission, on which also sat architect John Carrère, to beautify the city, beginning by grouping public buildings around a focal civic center. The commission's report, the "Group Plan" issued in 1903, proposed to clear 100 acres of densely built-up slums along the

Lake Erie shoreline and replace them with a parklike mall of double tree allées extending south from the lake based on the Place de la Concorde in Paris, to be lined by three-story government and civic buildings sharing a cornice line and white neoclassical columned facades. Much of it was realized in subsequent years.

Also in 1903, Burnham was retained by the Association for the Improvement and Adornment of San Francisco to transform it into "a more agreeable city in which to live." San Francisco had grown up willy-nilly after the Gold Rush, with no plan beyond an inflexible grid of streets laid over its sandy hills. It was rough-hewn and diverse, populated by immigrants from everywhere. And it had ambition—not merely to become a great city, but to become the Rome of the Pacific. Mayor James Phelan thought that if San Franciscans erected monuments and buildings like those of Athens under Pericles, they would be sure to "render the citizens cheerful, content, yielding, self-sacrificing [and] capable of enthusiasm." Burnham planned a formal neoclassical civic center at Market Street and Van Ness Avenue, from which other avenues radiated outward to squares marked with monuments. Prominent hilltops, like Telegraph Hill, would be made over into Roman visions complete with promenades, balustrades, fountains, and statues. Finished seven months before the devastating 1906 earthquake, Burnham's grand vision was shelved, though some elements of the effort were built: the Beaux Arts city hall, finished in 1915, with its huge dome, 42 feet taller than that of the US Capitol in Washington, the Dewey Monument at Union Square, and, at the corner of Market and Dolores streets (where it was moved in 1925 from its original location at Market Street and Van Ness Avenue), the California Volunteers Memorial, commissioned and dedicated by Phelan in April 1906, even as the city still smoldered around him, to honor the state's Philippine war dead.

While working on the San Francisco plan, Burnham was hired by the US government to prepare plans for Manila, in the US-occupied Philippines, and for the proposed summer capital at Baguio. His plan for Manila was familiar: facing the bay were huge axes and green spaces lined with solemn government buildings set wide apart, while a regular system of diagonals and squares ordered the street grids. The plan for Baguio, in the uplands north of Manila, looked incongruously like Versailles. Burnham, who was a fan of Rudyard Kipling and believed fully in the White Man's Burden, was confident that in the Manila plan he had discharged his duty to his own country, and the occupied one: "Manila may rightly hope to become an adequate expression of the destiny of the Filipino people as well as an enduring witness to the efficient services of America in the Philippine Islands." Never mind that the American occupation was bitterly resisted by the Filipino people for over a decade, with an appalling loss of life.

Burnham went on to the biggest plan of his career: a regional city plan for his hometown, Chicago, completed in 1909, with Edward Bennett. As depicted in impressive watercolor renderings commissioned from the artist Jules Guerin, the city would be centered on a huge plaza, lined with white colonnaded buildings, all focused on a circular city hall topped with a truly massive and tall dome. From the plaza, diagonal avenues departed, forming a diamond matrix cutting through the street grid before meeting a giant orbital parkway. The plan extended more than 60 miles from the city center, nearly encircling it with parks and greenbelts. The entire Lake Michigan shoreline was projected as parks, with a harbor centered on the civic plaza protected by breakwaters. Much of the plan was realized over the years: 20 of 24 miles of lakefront became parks; Union Station was completed in 1925; Wacker Drive was built, the first double-level boulevard anywhere; and several neoclassical buildings were constructed to anchor the city center: a new Art Institute, the Field Museum of Natural History, and Shedd Aquarium. While the grandiose central plaza and dome weren't built, the bird's-eye images of them and the overall plan were reproduced worldwide, becoming influential touchstones of a long tradition—one which only gained strength in the twentieth century.

Even as the White City was being built, the neoclassical revival was gaining traction elsewhere. In France, the École des Beaux-Arts with its baroque symmetries pushed aside the eclectic styles of mid-century. In Great Britain, this influence swept aside the medieval revival and Victorian eclectic styles, which had in their turn in the 1830s and '40s dethroned the austere classicism of the eighteenth century. The growth of the British Empire in India and Africa in the late Victorian and Edwardian eras saw a surge of "popular imperialism" at home, which seemed to see its own image in classicism. Since Rome, classical style had been Europe's idea of imperial style, and the adoption of baroque classicism beginning with Queen Victoria's 60-year Jubilee in 1897 was no exception. It was, after all, the style of Wren at the Royal Naval College at Greenwich and John Vanbrugh at Blenheim Palace, and would do perfectly when London needed to be made to look more like an imperial city. In 1901, Sir Aston Webb was commissioned to design a Victoria memorial and a ceremonial way from Buckingham Palace to Trafalgar Square. Completed in 1913, they hit all the notes that Burnham had played so well: a double-alléed Pall Mall from the baroque Admiralty Arch to the Victoria Memorial, with its rond point, and refronting Buckingham

Palace with a Beaux Arts facade. Finally, London looked French, or like America reimagined by a City Beautiful planner.

The Edwardians exported the mode to the ends of the empire. When Edwin Lutyens, whose early career had been as a leading Arts and Crafts architect building in medievalist styles, arrived in India in 1912 to lay out the new capital, New Delhi, it was neoclassical grandeur he was after, and he had clearly studied Baron Haussmann and Daniel Burnham. His plan, with its enormous, tree-lined Kingsway, bounded by the enormous India Gate, Viceroy's House, Parliament House, and Secretariat Buildings, looks like Burnham's Washington, DC, but with Indian motifs added to the architecture to satisfy the dictate of King George V, emperor of India, that the empire be based on the consent of the people.

George V was deluded, as history came to show. Empire builders tend to be. His resort to grandiose architecture to cover over the oppression and injustice that all empires rest on was far from the last in the twentieth century. The Italian dictator Benito Mussolini planned EUR (Esposizione Universale Roma) in Rome in the late 1930s as the site for a world's fair to be held in 1942 to celebrate 20 years of fascist rule. The fair was canceled because of World War II, but much was built: in neoclassical style, and in Rationalist style, a hybrid variant with a modernistic lack of decoration, but equally monolithic, white, symmetrical facades. In Nazi Germany, the late nineteenth-century neobaroque that Hitler, a wannabe architect, admired as a boy was the basis of his program to build massive spaces for public spectacles—urban stage sets, not unlike sinister versions of the Chicago fair. In Munich was the Haus der Deutschen Kunst (House of German Art), by Paul Troost. In Nuremberg was the Congress Hall, modeled on the Roman Colosseum, by Ludwig Ruff, and the Zeppelinfeld rally grounds (made famous in Leni Riefenstahl's propaganda film *Triumph of the Will*), by Hitler's favorite architect, Albert Speer. Speer conceived a monumental intervention into the heart of Berlin to symbolize its future status as the Welthauptstadt Germania (World Capital City of Germania). It was to be anchored by the Volkshalle (People's Hall), a gigantic pastiche of Hadrian's Pantheon in Rome and the Paris copy of it, which would have been 950 feet high, topped with a dome 820 feet in diameter, big enough to accommodate 180,000 people. It faced down a five-kilometer-long parade ground on the north-south axis, which passed under a triumphal arch so large that the Paris Arc de Triomphe could fit inside its opening.

Architecture itself is not evil or good; it has no intrinsic qualities. Only the intentions of its makers, and the actions of its users, impart a moral valence. Speer's Berlin was no different in architectural terms than Hadrian's Rome or Lutyens's New Delhi, just much, much bigger. But how cities are designed to act, shape, limit, control, or obscure the lives of their inhabitants is a question with moral dimensions. Speer's plan shows how architecture can oppress—potentially by shaping and limiting space, but also by hiding it: it can be deployed as a screen to obscure realities inconvenient to those in power. The neoclassical is at its base always a mask, as it has no real history, no real place, no real context. And yet by resembling the past, it claims continuity, which can be read as a claim to legitimacy for present regimes. This is why so many rising but insecure states and societies are drawn to it, and why totalitarians in particular have found it irresistible. Joseph Stalin, provincial that he was, was enchanted by an especially florid vein of neobaroque neoclassicism, and saw to it that it was built throughout the Soviet world, in a dizzying range of variants that, if all brought together, would resemble at monumental scale the display cases in a Viennese cake shop. The Red Army Theater and All-Russian Exhibition Center in Moscow are just the tip of the mountain of confectioner's sugar. The Soviets even exported the style to China, building Tiananmen Square, the largest public square in the world, flanked by massive columned buildings in the Stalinist neoclassical mode.

City builders in the democratic world have also been drawn to City Beautiful ideas and the methods of neoclassical design, if not the decoration of neoclassical buildings, to give grandeur, order, scale—and to some eyes, beauty—to large urban spaces. Canberra, federal Australia's new capital city, was laid out in 1912–13 by American architect Walter Burley Griffin (who, as a young man in Chicago, was an admirer of Louis Sullivan and an employee of Frank Lloyd Wright) in a combination of Burnham's diagonal avenues and squares system, Frederick Law Olmsted's picturesque parks and lakes, and a set of connected, suburban circles à la Ebenezer Howard's garden city. The Swiss architect Le Corbusier based his modernist urban-planning schemes, the Plan Voisin and Radiant City, on regular, geometrical plans that are directly in line with the mode (see chapter 3). He even talked, as Daniel Burnham did, of the power of the beauty of a regular plan to motivate the citizens to feel pride in their environment, and to live orderly, rational lives. When he was given the chance to actually design and build a city—Chandigarh, the new, postindependence capital of the Indian state of Punjab, which he planned in the 1950s—he did so on a geometrical matrix directly descended from Lutyens, Burnham, and Haussmann.

On a more limited scale, modernist architects in the United States used neoclassical methods to determine the relationships between buildings and public space, especially in civic centers and museum districts, using symmetrical, colonnaded facades facing one another across plazas centered on fountains to confer grandeur and gravity. Lincoln Center in New York, the Kennedy Center in Washington, the Music Center in Los Angeles, and the Illinois Institute of Technology campus in Chicago are examples, each essentially neoclassical. There are countless others like them.

Of late, different architectural flavors have obscured the common parentage. Frank Gehry's iconic, post-postmodern civic buildings, like the Guggenheim Museum in Bilbao and the Walt Disney Concert Hall in Los Angeles, are scaled and sited according to the same logic as Burnham's Court of Honor at the Chicago fair or the Victoria Memorial in London: they are monuments, meant to manifest and direct glory and gravity to their cities and those who preside over them. They testify to the continuing need of civic and national elites to assert their virtue through visible investments in physical form, and to the continued success of the enterprise. The so-called Bilbao Effect—when a single, highly visible building and its strategic placement in the urban fabric exerts a transformative effect on that city's economy and perceived status—has become an accepted truism for how cities can put themselves on the map. No less than a triumphal arch, a trophy building of this kind, sited correctly by a skilled architect or planner, seems to change everything. And it's much cheaper to write a big check for a big building than to actually change the world.

2. Monuments

Field Guide: Neoclassicism

Diagnostics:

- Buildings: Neoclassical architecture, usually white, symmetrical, with classical columns, pediments, friezes, moldings, etc. Many possible variants, including Beaux Arts mixes of different genres, and modernist versions with regular, white columns and symmetrical facades.
- Streets, boulevards, and parks: straight, lined with allées of trees; "axial"—originate from buildings at fixed angles and often terminate in circles or plazas.
- Monuments: figurative sculptures, usually in marble or bronze, monochrome, with a classical feel; usually celebrating (dead) national heroes.

Examples:

USA

- Most state capitol buildings.
- New York: New York Public Library, Metropolitan Museum of Art, Grant's Tomb, Grand Army Plaza.
- Washington, DC: US Supreme Court, Capitol Building, White House, Capitol Mall, Lincoln Memorial, Jefferson Memorial.
- Richmond, Virginia: Monument Avenue.
- Philadelphia: Benjamin Franklin Parkway Museum District, Philadelphia City Hall, Philadelphia Museum of Art.
- Pittsburgh: Schenley Farms district.
- San Francisco: Civic Center.
- Wilmington, Delaware: Rodney Square.
- Denver: Civic Center, Greek amphitheater, Voorhies Memorial, and the Colonnade of Civic Benefactors (1919).
- Coral Gables, Florida: City plan.
- Sarasota, Florida: John Nolen's 1925 city plan.

United Kingdom

- London: Trafalgar Square, Admiralty Arch, Pall Mall, Victoria Memorial.

France

- Paris: Arc de Triomphe, Haussmann's Grands Boulevards.

Austria

- Vienna: Ringstrasse.

Australia

- Canberra: Australian Capital Territory plan.

South Africa

- Cape Town: Rhodes Memorial.
- Pretoria: Union buildings.

China

- Beijing: Tiananmen Square.

India

- New Delhi: Capital district plan, Kingsway, India Gate, Viceroy's House, Parliament House, and Secretariat.
- Calcutta: Victoria Memorial.
- Bombay: Town Hall.
- Hyderabad: British residency.

Modernist Variants:

- New York: Lincoln Center.
- Los Angeles: Music Center.
- Chicago: Illinois Institute of Technology.
- Washington, DC: Kennedy Center.

Italy

- Rome: EUR (Esposizione Universale Roma).

Brazil

- Brasília city plan.

Images:

Palace of Mechanic Arts, World's Columbian Exposition (1893), Chicago, Illinois.

Court of Honor, World's Columbian Exposition (1893), Chicago, Illinois.

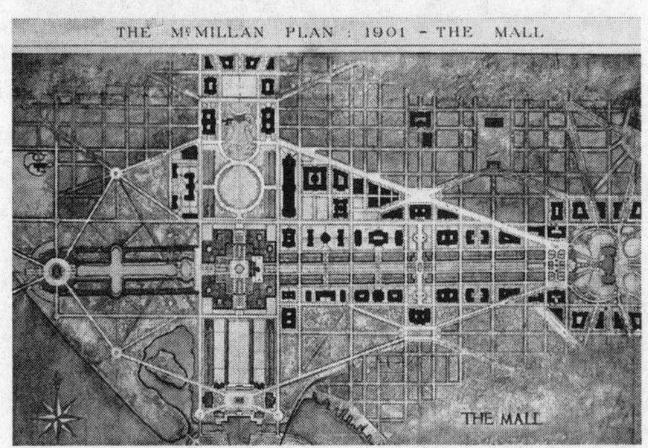

McMillan Plan for the National Mall, Washington, DC (1901). Plan: Daniel Burnham, Frederick Law Olmsted, Jr., and Charles McKim.

New York Public Library, New York, New York (1902). Architects: Carrère and Hastings.

Plan of Chicago, Civic Center (1906). Architects: Daniel Burnham and Edward Herbert Bennett. Rendering: Jules Guerin (1908). *Typ970U Ref 09.296, Houghton Library, Harvard University*

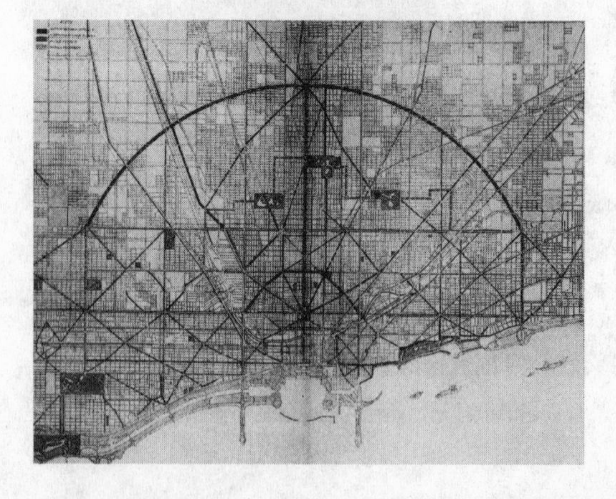

Plan of Chicago (1906). Architects: Daniel Burnham and Edward Herbert Bennett.

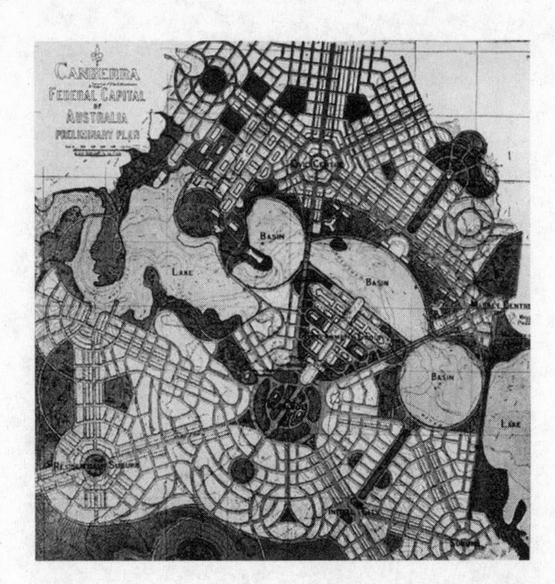

Australian Capital Territory plan (1913), Canberra. Plan: Walter Burley Griffin.

Metropolitan Opera House, Lincoln Center for the Performing Arts (1966), New York, New York. Architect: Wallace Harrison.
Luigi Novi

Chapter [3]

Slabs

Le Corbusier, Robert Moses, and the Rational City

A city is an attempt at a kind of collective immortality—we die, but we hope our city's forms and structures will live on. Ironically, our attachment to these forms makes us more vulnerable than ever: now there are more ways than ever for our lives to be destroyed.

—MARSHALL BERMAN, "FALLING" IN *RESTLESS CITIES*

"This new doctrine of modernity . . . is inspired by the self-delusion we found in Le Corbusier that there is a new thing called modern man, a new animal without roots in the past and whose mind is, or must become, a tabula rasa.*"*

—DAVID WATKIN, *MORALITY AND ARCHITECTURE REVISITED*

You can walk around New York, like most big cities in the world, and not see them, though in large parts of the city they are ubiquitous, rising straight up like enormous, grid-covered boxes in brick, concrete, or glass, standing alone or stretching for blocks, with an undeniable resemblance to gravestones: *slabs.* They are usually set back from the street, surrounded by parking lots, fenced-in lawns, and confused amalgams of sidewalks, planting, and pavement, arrangements that might resemble parks but are clearly no such thing; or they sit on or hover above sprawling, hard, windswept plazas, whose intended purposes are unclear. They may house business offices or people in apartment flats. Whether those residents are poor, middling, or rich can be determined only by the quality or lack of upkeep of the spaces between the building and the street—the architectural form of the slabs is generally the same. Either way, they are unwelcoming, evidently designed to be separate from those walking past on the city street. In New York City, big chunks of all five boroughs are given over to series and clusters of slabs set apart in superblocks—extra-large blocks where the intervening crossing streets have been closed off and replaced with flat areas dedicated to grass, trees, concrete, roads, or parking. These slabs may be private buildings guarded by doormen and security companies, or public housing projects pointedly left to decay, guarded by opposing pickets of street gangs and police. They may house banks, insurance companies, or hotels. They seem to speak of modernity, yet they can't avoid testifying to the failure of the twentieth century to live up to its urban dreams, because whether they look prosperous or impoverished they point to the unbridged divide between the two that afflicts the modern world. They may be in New York or Santa Monica, Tokyo or Caracas. Once you start to see them, you will see that they are everywhere, in almost every major city on this planet. The question is, why?

The sheer ubiquity of slabs would argue for an almost mechanical origin, as if they were something inevitable and preordained, arising spontaneously with the advent of modernity, somehow related to the mysterious concrete slab that keeps cropping up in the movie *2001: A Space Odyssey*. Building slabs has been a collective exercise, or at least a widespread one, for more than a century. Many tens of thousands of different agents—architects, planners, developers, governments—have built them, to supply new housing for growing populations, and to accommodate new modes of transportation, especially cars. In this sense, slabs were a "rational" choice made by many people in many places, in response to actual modern conditions.

But they were not the only choice, just one among many. Their proliferation attests to common desires as much as common needs, namely to participate in the modern world by adopting modern building techniques and forms. At the root of their appeal is the fact that slabs come from a utopian desire, as strong as any in the utopian canon, conceived to save the world from its urban mistakes. And, remarkably, their spread is in good part traceable to the influence of a single man, by turns charismatic and curmudgeonly, prophetic to some, but dubious to others, whose unlikely vision shaped our world.

Charles-Édouard Jeanneret-Gris was born October 6, 1887, into a prosperous family in the town of La Chaux-de-Fonds, in the Swiss canton of Neuchâtel in the Jura Mountains, about five miles from the French border. La Chaux was French speaking, heavily Protestant, counting many descendants of Cathar refugees from the South of France, and bourgeois par excellence. Karl Marx, who visited, called it "one unified watchmaking industry." The compact town consisted of factories, ateliers, banks, cheap rental apartments for workers, and luxury houses for the business class. Its culture reflected its economic underpinnings: a respectable regional-syndicalist business politics at once conservative and mildly communitarian, and a strong influence of Freemasonry, then much in vogue in the francophone world, attractive for its blend of watered-down Christian mysticism, occult rites and initiations, equation of spirituality with geometric forms, and focus on moral improvement. Charles-Édouard was exposed early to the town's masonism, as the Loge l'Amitié, based on the Scottish Reformed Rite, which interestingly posited the existence of a Grand Architect of the Universe, was popular among the Jeanneret family's social set, and the lodge's charity activities were highly visible: a library, nursery schools, education, and food programs for the poor. In spite of a suggestive index card held in the files of the World War II Vichy French secret police, it is unknown whether Charles-Édouard ever became formally initiated. But Freemasonry would hold a lasting influence on his work and philosophy.

He attended the École d'Art in the town, under several teachers conversant with the new trends in art and architecture then in vogue elsewhere on the Continent. At a young age, he began to receive architectural commissions, for residences. Most of these residences were indebted to European Arts and Crafts variants, including the Vienna Workshop and the German Crafts movement, such as the 1912 Villa Jeanneret-Perret influenced by John Ruskin, who had once sojourned in La Chaux. At least one residence integrated consciously "Swiss" rooflines. Another, designed when Charles-Édouard was 17 years old, was neoclassical, with a tall, rectangular block adjoining a semicylinder, wrapped with cornices and incongruously decorated with dainty oval windows and thin first-floor columns, as if an early example of the 1940s Hollywood Regency style.

During the Great War Charles-Édouard stayed in neutral Switzerland, teaching at the École and, like many architects of his generation, making plans to rebuild areas damaged in the conflict. In 1914, to help reconstruct Flanders, he drew a concept for a "Domino" building frame, consisting of concrete slab floors supported by steel-reinforced concrete columns set inboard of the slab edges, thus making the structural support system independent of the exterior cladding, freeing it, for example, for long bands of windows. Jeanneret's ideas were clearly influenced by engineering innovations in American industrial construction being publicized in Europe as well as the versions of them emanating from more highbrow European faculties such as Germany's Bauhaus.

To expand his education, Charles-Édouard set off on a trip, visiting Vienna, where he read the modernist pioneer architect Adolf Loos, and Greece, where he spent 18 days with the monks of the Orthodox monastery atop Mount Athos, impressed by the spectacular site and the austere monastic architecture and lifestyle. Exposure to the new avant-garde movements led Jeanneret to turn away from the nineteenth-century modes he had been taught. When he returned to La Chaux, it seems that these new convictions, combined with business difficulties and personality conflicts with his former teachers and associates, convinced him to make a fresh start. In 1917, he moved to Paris, taking lodgings at 20 Rue Jacob, Saint-Germain-des-Près, and fell in with a group of expatriate Swiss bankers and other young men attracted to the energy and new ideas of the French capital.

In Paris, Charles-Édouard Jeanneret-Gris, the provincial Swiss bourgeois, began to transform himself into something new. He met the painter Amédée Ozenfant, who practiced what he termed "Purism," a more straightened derivative of the Cubism of Léger and Picasso. The two started a small magazine, *L'Esprit nouveau*, in which they presented uncompromising avant-garde notions of art and life. Jeanneret began to sign his contributions with pseudonyms, including, beginning in 1920, "Le Corbusier," a name perhaps derived from his distant Southern French ancestors, perhaps simply made up, and suggestive of the word for crow, *corbeau*; it had the added quality of referring to himself in the third person. It was apparent to many around him that the young Jeanneret had grand plans, which required leaving La Chaux-de-Fonds behind. When he left, a friend wrote a novel with a character based on him: "He is unloading ballast. That is how you rise."

Jeanneret later wrote to another friend:

> *LC is a pseudonym. LC creates architecture, recklessly. He pursues disinterested ideas; he does not wish to compromise himself in betrayals, in compromises. It is an entity free from the burdens of carnality. He must (but will he succeed?) never disappoint. Charles-Édouard Jeanneret is the embodied person who has endured the innumerable radiant or wretched episodes of an adventurous life. Jeanneret Ch. E. also paints, because, though not a painter, he has always been passionate about painting and always has painted—as amateur.*
>
> *Ch. E. Jeanneret and Le Corbusier both sign this note together.*
>
> *Warmest regards,*
>
> PARIS 18 JANUARY 1826.

The new man was attracted to uncompromising, idealistic notions. Perhaps influenced by the austere Calvinism of his hometown, perhaps by Freemasonry's fixation on geometric and abstract concepts of spiritual and moral order, he set out to define for himself a destiny worthy of his new identity. He wrote a "Poem of the Right Angle," a sort of manifesto of the formal clarity and moral and spiritual righteousness of the 90-degree angle and of the straight line. He proposed "The Modulor," a system of proportion based on the "eternal harmony" of the human body—rehashing the anthropometric ideas of the Greeks and the Roman architect Vitruvius. One of Le Corbusier's biographers, Charles Jencks, wrote of him that "he would next try to find a heroic role in identifying his personal destiny with that of a vast impersonal force," and that "throughout his life, Corbusier was searching for a type of universal symbolism that would be trans-historical and non-conventional." Le Corbusier wrote, in a Nietzschean mode, "I want to fight with the truth itself."

In the 1920s, he received several commissions for private houses, mostly from acquaintances, some Swiss-connected bankers and industrialists. The structures he designed were white and aggressively austere, with rounded stairwells, tube railings, and horizontal strips of windows—all derived from the industrial-minimal aesthetic of the German Bauhaus and others, combined in a way that would become the architect's signature style. In 1922, he was invited to submit a house design for the Paris Salon d'Automne exhibition. In addition, he was asked if he might provide some ideas for outdoor urban design in Paris. What he supplied was astonishing: a "Plan for a Contemporary City of 3 Million Inhabitants," in the form of a 100-square-meter diorama of an ultramodernist city. The Ville Contemporaine featured a grid of identical 60-story, cruciform, steel-and-glass towers, each separated by 400 meters from the next, all set in a parklike composition of green space. In the center was a multimodal transportation center, and a corridor of grade-separated roads differentiated according to the type of traffic each would carry, and segregated pedestrian decks. On one end was an airport. The tower grid was surrounded by repeating arrays of smaller, 6-story buildings in zigzag shape, called "villas," also set in green space and ringed by one-way viaducts. The architect's guiding principle was separation: of different functions of the city—industrial, commercial, residential, and recreational—from one another, of different classes and speeds of traffic, and of buildings from the ground plane; the building footprints would occupy only 85 percent of the surface area, leaving the rest for circulation and what he called "gardens."

This applied to classes of people as well: the towers would house businesses and the wealthier residents; the villas were for the workers. There was no town hall, courthouse, or cathedral.

It was clear that postwar Paris faced serious crowding, congestion, and deteriorating buildings and infrastructure, and desperately needed new housing, offices, and transportation ideas. But Le Corbusier's Contemporary City represented a complete rejection of Paris as it was: a many-layered city accreted around its medieval core, its blocks and streets irregular and filled with a tumult of people, vehicles, and different kinds of activities around the clock. Le Corbusier disliked this Paris, like so many other critics of unplanned, traditional cities then and for centuries before, equating their mixing and heterogeneity with an archaic disorder. "The lack of order to be found everywhere in them offends us; their degradation wounds our self-esteem and humiliates our sense of dignity. They are not worthy of the age; they are no longer worthy of us," he wrote. In particular, he hated the traditional street: "It is the street of the pedestrian of a thousand years ago, it is a relic of the centuries; it is a nonfunctioning, an obsolete organ. The street wears us out. It is altogether disgusting! Then why does it still exist?" He would do away with all of it and substitute a rational, ordered, and rectilinear—because, he asserted, "the straight line is the line of man, the curved line that of the donkey"—city matrix fit for the new society and the "new man" of the machine age.

There can be no doubt that Le Corbusier wanted his city to be perceived as visionary and radical, but in most senses it was already old hat. The towers were nothing new: America's skyscrapers had captivated people around the world for decades; Louis Sullivan, the Chicago pioneer of steel-framed skyscrapers, had himself proposed cruciform towers in 1890. Nor was reinforced concrete. The French architect Auguste Perret had designed tall, reinforced concrete buildings as early as 1903–4, with bays he called "rédents," a word Le Corbusier also used. Le Corbusier was doubtless aware of the work of the Italian Futurists, most notably Antonio Sant'Elia's 1914 "Cittá Nuova" plan for a Milan rail station with multiple levels of grade-separated transport, itself inspired by a long line of American visions of vertical, modern cities serviced by new inventions like dirigibles. The separation of traffic modes was an idea dating back at least to 1860 and by the 1920s was a staple of popular fiction and illustration. Big cities had subways and parkways, especially in America. European architects were eager to adopt them, too. For Berlin, Hans Poelzig proposed a Y-shaped skyscraper for Friedrichstrasse in 1921, and Ludwig Hilberseimer drew his High Rise City in 1924, with slabs set in rows and vehicle and pedestrian traffic below on separated roads and elevated walkways.

In France, the inescapable precursor was the architect Tony Garnier, who exhibited plans in 1904 for Une Cité Industrielle, a planned industrial city for 35,000 people organized on the principle of separating different functions in space, amounting to one of the earliest forms of zoning. Le Corbusier admired it, writing, "one experiences here the beneficent results of order. Where order reigns, well-being begins." On one hand this statement sounds distinctly like the fascist language of the era; on another, it evokes the dogmas of contemporary socialists. It reflects a belief that design can make us well, or better, like a kind of medicine of concrete and steel, prescribed by a doctor, the architect. This faith was widespread in the profession then. Garnier had been influenced by the nineteenth-century novelist Émile Zola's utopian book *Travail*, just as countless architects of rationalized, utopian communities had been influenced by the beliefs of Robert Owen and Charles Fourier that the new industrial rationality could remake society for the better. Owen had called his brick quadrangles "moral quadrilaterals," with on one side a model factory, another a communal dining room, another meeting rooms for recreation, and another apartments. Fourier's phalanstery contained theaters, gardens, promenades, and (being French) "gourmet cuisine for everyone," but it was essentially a highly ordered factory-city that housed 1,620 people in one building.

Le Corbusier's vision was also clearly in the long tradition of massive, state-decreed building and rebuilding projects in France, from Louis XIV's Versailles to Baron Haussmann's violent transformations of Paris in the mid-nineteenth century. Le Corbusier would have known of the work of Eugène Hénard, the Beaux Arts architect who directed the continuation of Haussmann's *grands travaux* in Paris in the Gilded Age, helped plan Paris's 1889 and 1900 world's fairs, and published eight studies for classicist City Beautiful transformations of the city between 1903 and 1906. With its avenues, axes, and grand arteries, the Ville Contemporaine was no less an exercise in classicist City Beautiful planning, though with a stridently anti–Beaux Arts modernist surface. What can be said to have been different was, oddly, an uncompromising form of Americanism. US skyscrapers had to that point universally downplayed their innovative modern structures by cloaking them in historicist detail: Gothic or classical, with elaborate cornices, sculpture, and ornament of all kinds. Le Corbusier ridiculed this backsliding, instead looking to the industrial rationality below the skin: "Let us listen to the counsels of American engineers. But let us beware of American architects." His slabs were white, featureless, and utilitarian, like the factories organized by Taylorist and Fordist principles of maximum efficiency—doctrines that inspired him.

What was truly remarkable about the plan was that Le Corbusier promised to give people everything (he thought) they wanted: both the city and the country, in one package. Paradoxically, he insisted that more concentration, not dispersal, was the cure to urban ills—by stacking people in dense towers separated by green space. In essence, he would resolve the tension between city and country by mating them, the result being "towers in a park," as his prescription became known. "We must increase the open spaces and decrease the distances to be covered. Therefore the center of the city must be constructed vertically," he wrote. His plan was hyperurban—organized around business and enabled by fast new technologies of transport, the exact technologies that were making suburban dispersal possible. Echoing the Futurists and their worship of *velocittá* (speed), Le Corbusier would later write: "A city made for speed is a city made for success." At the same time it preserved the old arcadian dream of escaping the city to a picturesque green nature, by stretching that nature between the buildings and under the highways. Thus the New Man in the New Society could have gleaming American skyscrapers, nestled reassuringly in extensive French parks. Le Corbusier much admired Versailles, that great model of the orderly clockwork universe as a diagram of its designer's omniscient control. He frequently reproduced images of Paris's Tuileries and Jardin du Luxembourg in his books as reference points for his own plans and was powerfully influenced by the picturesque Bois de Boulogne and Parc Monceau, designed by the eighteenth-century French architect Francois-Joseph Bélanger (1744–1818).

It was a bold and brilliant sleight of hand: *Voilà! Tous les deux à la même fois*. Both at the same time. All the advantages of modernity and none of its disadvantages. Le Corbusier called this, and other pivotal moments in his career, "a flash of intuitive insight." It is fair to say these weren't flashes of originality so much as instances of his uncanny ability to glean aspects of things he saw around him, incorporate them on a new canvas to appeal to new sensibilities, and triumphantly sign his name to them. The Contemporary City though remained an odd amalgam of contrary principles, not entirely coherent or comprehensible. It was based on the straight line, yet completely dependent on the picturesque, that is, curving, view of nature. It demanded speed in order to slow down the pace of urban life and decrease commuting. It required extraordinary densities to preserve open space. Indeed, in a fundamental sense, Le Corbusier was proposing incommensurable things, yet this didn't dim the appeal of his vision to numerous people, then and now.

Le Corbusier's Ville Contemporaine wasn't built, but the architect's reputation as a radical modernist visionary continued to grow. In 1923, he published a manifesto entitled *Vers une architecture*, published in English as *Toward an Architecture*, which, according to the British critic Reyner Banham, "was to prove to be one of the most influential, widely read and least understood of all the architectural writings of the 20th century." It has enthralled generations of architecture students owing to its paradigmatically modernist machine aesthetic and definitive program, which Le Corbusier broke down into five principles: buildings raised on columns (pilotis); roof gardens; "free" interior plans, without dividing walls; horizontal strip windows; and "free" facades, without structural or ornamental texture. It has been interpreted in widely divergent ways by different readers, in part because of the maddening nature of Le Corbusier's writing—a collage of images, text, diagrams, and aphorisms, deployed in 50 books published in his lifetime. Charles Jencks wrote that Le Corbusier's writing "had a hypnotic effect," in spite of or because of this form: "short chapters introduced by an 'argument,' in blank verse, which is reiterated throughout the text so persuasively that one forgets to note both its dubiety and illogicality."

In 1925, Le Corbusier displayed another, newer city model at the Paris International Exhibition of the Arts, this time not abstract but alarmingly concrete: he proposed to go beyond Haussmann and rip out much of the historical heart of Paris and replace it with a regimented, rationalized, purist version of a distinctly American-looking future. Named the Plan Voisin after the French automaker Gabriel Voisin agreed to sponsor it (Peugeot and Citroën had demurred), the plan called for the demolition of a two-mile swath of the historic heart of the Right Bank and its replacement by a grid of 50 or more superblocks with 26 soaring office towers. As in the earlier Contemporary City, these were huge cruciform towers, but this time they were 60 stories high, separated by 400 meters, and completely covered in glass. The business center, with offices for "400,000 clerks," was strictly separate from the workers' housing, not in "villas," a holdover from the garden city tradition, but in linked, linear slabs. All the buildings were raised on pilotis, allowing 95 percent of the ground in the business district and 85 percent in residential areas to remain open. Underground stations under each tower linked to a seven-level system of main train lines, suburban lines, subways, highways, an airport, and elevated pedestrian decks and walkways. The entire plan was threaded with fast, separated roads, making the car the king of the future. The architect rhapsodized: "Those hanging gardens of Semiramis, the triple tiers of terraces, are 'streets of quietude.' Their delicate horizontal lines span the intervals between the huge vertical piles of glass, binding them together with an attenuated web. Look over there! That stupendous colonnade which disappears into the horizon as a vanishing thread is an elevated one-way autostrada on which cars cross Paris at lightning speed. For twenty kilometers the undeviating diagonal of this viaduct is borne aloft on pairs of slender stanchions." At last, "those gloomy clefts of streets which disgrace our towns" would be wiped away, as surely as Baron Haussmann wiped away the medieval quarters of Paris. And, Le Corbusier concluded, "then the street as we know it will cease to exist."

Despite his best efforts, he found no customers to fund his urban vision. For years, he appealed to capitalists (after all, his Swiss friends in Paris were bankers), asserting that the increase in land values due to higher densities would justify the enormous expense of leveling much of the capital city. "To urbanize is to valorize," he claimed. "To urbanize is not to spend money, but to earn money, to make money." When this failed to interest investors, he appealed to the French authorities: "The idea of realizing [this urban vision] in the heart of Paris is no utopian flight of fancy. There are cold figures to substantiate this thesis. The enormous increase of land-values that must result would yield a profit to the state running into billions of francs—for to acquire the central part of Paris and redevelop it in accordance with a coordinated plan means the creation of an immense fresh source of wealth."

Yet it was likely hard for the bankers or the bureaucrats to ignore the more alarming implications of his program, which was, first of all, predicated on the massive demolition of the world's existing cities: "Therefore my settled opinion, which is quite a dispassionate one, is that the centres of our great cities must be pulled down and rebuilt, and that the wretched existing belts of suburbs must be abolished and carried further out; on their sites we must constitute, stage by stage, a protected and open zone." Secondly, the means required to effect it were, if not revolutionary, then certainly authoritarian. He prescribed three main "DECISIONS" upon which the plan would be executed: "1. Requisitioning of land for the public good. 2. Take an inventory of our cities' populations: differentiation, classification, reassignment, transplantation, intervention, etc. 3. Establish a plan for producing permissible goods; to forbid with stoic firmness all useless products. To employ the forces liberated by this means in the rebuilding of the city and the whole country."

Frustrated, Le Corbusier turned to the Soviet Union, designing a similar scheme for Moscow he called La Ville Radieuse, the Radiant City, which he exhibited in Brussels in 1930 and published in book form in 1933. The book was dedicated "To Authority," in a clear indication that he understood the level of destructive power that would be required to realize his dreams. The high-rise housing was based on the allocation of precisely 14 square meters per occupant—a standard cell, like those he had admired at the Mount Athos monastery. Though in his writings he pointedly criticized Western garden city ideas and reviled American suburbia for its complicity in "the organized slavery of capitalist society," the Soviet authorities didn't bite, either. It didn't help that they were already seduced by the idea of dispersal: radically decentralizing people away from cities into something like garden suburbs. Le Corbusier received a single Soviet commission: the Centrosoyuz consumer union headquarters, in Moscow, a modest midrise building, its first floor raised on pilotis, built during 1928–36. In a show of his ideological malleability, he even appealed to the fascist French Vichy authorities during the World War II, again to no avail.

In the difficult environment of the late 1920s and '30s, he did gain some commissions, mostly for suburban villas: the 1932 La Roche Jeanneret house in Paris, with a roof garden; the 1932 project for an apartment house in Zurich (unbuilt); and the Villa Savoye at Poissy, outside the French capital (1929–31). Le Corbusier's famous pronouncement that a house was "a machine to live in" attached most in the public mind to this exaggeratedly machinelike building, which became a twentieth-century modernist icon. It was starkly antinature: elevated off the ground, with sealed windows, asserting its independence of the earth like a spaceship. Yet it depended on picturesque effects: views over unpeopled, green space, a kind of virginal isolation. On one hand Le Corbusier would seem ungrateful, if not dishonest, in selling his Machine Age on the back of a romanticized nature. On the other, he understood that this paradox was irresistible to a growing group of people who wanted to live in a "modern" way, in a kind of temple of technology, cleanliness, and wealth, and yet luxuriate in pastoral splendor, separated as far as possible from any sign of other human beings.

It is interesting to note, as the scholar Sven Birksted has pointed out, that Le Corbusier likely took his inspiration for the villa from an eighteenth-century one by Bélanger for the comte d'Artois. A contemporary sketch shows the resemblance: a white villa with a flat roof, square plan, and white columns sitting atop a green hill in a picturesque landscape complete with a foxhunt. Similarly, evidence points to Le Corbusier's copying a number of his uncompromising modernist buildings from the works of Bélanger, a late-eighteenth-century classicist with romantic tendencies whose books he carried with him throughout his life—a parentage that Le Corbusier did little to acknowledge. He also seems to have borrowed from Bélanger the use of Freemasonic symbolism, a vocabulary of forms aspiring to universal, abstract meaning that allowed Le Corbusier to bridge otherwise irreconcilable categories. He could compare his skyscraper mass housing schemes to the Parthenon in Athens, and his suburban villas for industrialists to ancient temples. Ever opportunistic in seeking patronage, he used combinatory formulae to be equally "polyvalent, malleable, and passe-partout," according to Birksted: "This symbology . . . allowed the ne plus ultra of true originality: mystery and indecipherability."

During the difficult period from the economic crash of 1929 through the war, Le Corbusier busied himself drawing hypothetical city plans—the polar opposite of his suburban villa, with no attempt to integrate or reconcile their divergent premises. In 1929, he traveled to South America, ever in search of commissions. His schemes evolved. Riding in planes, he was impressed by the variety of regional topography he saw; he compared the curving landforms of Rio de Janeiro to a woman. His studies for a plan for Buenos Aires took into account the city's hills and location on the River Plate—hardly revolutionary but a step forward from his earlier flattened, hypothetical schemes. Plans for Rio, Buenos Aires, and Montevideo show major arterial roadways following shorelines—curving, like the donkey's path—and his repeating skyscrapers separated by irregular intervals. He was loosening up a little. For four years he worked on ideas for Algiers, the skyscrapers evolving into linear series on asymmetrical plans, and changing shape into H's, Y's, and lozenges. He made plans for Stockholm, Nemours, Bogotá, Moscow, Izmir, and other cities. Charles Jencks summed up: "His output of city plans is remarkable, not only in sheer size, but also in terms of futility. Few were commissioned, fewer still were paid for and perhaps none stood the slightest chance of being adopted. This may account for the new tone which is discernible in

Le Corbusier's writings: diffuse, repetitious, sometimes bombastic and always now in an extreme hurry."

Le Corbusier kept busy partly by lobbying his own profession. In June 1928, he gathered 27 other modernist architects in a borrowed Swiss castle and founded CIAM: Congrés internationaux d'architecture moderne (International Congresses of Modern Architecture). Gaining new adherents, the group also met in 1929, 1930, and 1932, plotting modern architecture's role not only in constructing buildings but in redesigning the economic and political underpinnings of society. The fourth meeting, in 1933, which had been planned for Moscow, a location that was scotched after Le Corbusier failed to win the competition for the projected Palace of the Soviets, was instead held aboard a ship sailing from Marseilles to Athens. Dominated by his agenda, the group discussed the separation of cities into four "functional" zones: for dwellings, work, recreation, and transportation. Though the idea wasn't published until a decade later, by Le Corbusier acting alone, the "Athens Charter" principles of functional segregation in city planning effectively became attached to International Style architecture as an obligate single dose of the modernist prescription for the world's ills.

In the fall of 1935, Le Corbusier spent nearly two months in the United States, lecturing across the East and Midwest, his tour sponsored by the Museum of Modern Art (MoMA) in New York, which had included his work in its 1932 *Modern Architecture* show. Forty-eight years old, he had long incorporated preconceptions about America into his work, before ever setting foot there. In his earlier writings he had extolled American industrialism, sprinkling his books with images of silos and factories. But he withheld praise for America's greatest cities, declaring them too chaotic, a natural reflection of American capitalism's disorder and destructiveness. Of New York, which he called "a barbarian city," he wrote in 1923: "As for beauty, there is none at all. There is only confusion, chaos and upheaval." On his actual arrival he reported, satisfied, "Yes, the cancer is in good health," deploying his favorite biological metaphor for cities. He took pleasure in provocative rhetoric, calling New York and Chicago "mighty storms, tornadoes, cataclysms . . . so utterly devoid of harmony." The world-famous skyscrapers disappointed him in person: they were too close together, and too short. Two days before his inaugural lecture at MoMA on October 24, the *Herald Tribune* printed this headline over its interview with him: "Skyscrapers Not Big Enough, Says Lecorbusier at First Sight. French Architect, Here to Preach His Vision of 'Town of Happy Light,' Thinks They Should Be Huge and Farther Apart." He drew diagrams showing New York's evolution: first gradually being built up in its current, chaotic form, then "tomorrow" transformed into a Radiant City with huge tower slabs spaced widely apart. He preached his gospel of separation and speed, and sketched a future utopia that depended on demolition of existing cities, which he assured his audiences were "temporary."

Upon his departure for New York, he had every expectation of being offered commissions in the United States. He had American clients in France, after all. He assiduously searched out power brokers in New York and elsewhere who might make his schemes possible: mayors, officials of the New York City Housing Authority (NYCHA), and wealthy men such as Nelson Rockefeller. The need for a solution to the problem of America's inner cities was not in doubt. Americans' concern about the slums had been growing for decades; but the remedies most Americans favored framed the question in terms of buildings (the hardware), not in terms of people, jobs, wages, and the economy (the software). Jacob Riis's view was typical: "The poor we shall always have with us, but the slum we need not have." Build better houses for slum-dwellers, one Ohio official told Congress, and you will "do away with the slums."

Americans' prevailing attitude would seem to be receptive to Le Corbusier's pitch, but the architect faced multiple obstacles to realizing his urban vision in the United States. For one, he was foreign, and in the Depression the consensus was that available jobs should go to citizens. Second, his vision was widely perceived as being too modernist, and too European; what little public housing being built still looked traditional, like the NYCHA's Harlem River Houses, low brick tower blocks that Le Corbusier saw under construction. Third, the huge scale of destruction and construction required was impractical under the economic circumstances, to say nothing of grandiose. Fourth, the American housing reform movement was dominated by decentralists, such as Henry Wright, Lewis Mumford, and Catherine Bauer, who believed that slum-dwellers ought to be rehoused in planned communities built on cheap land on the periphery, not on expensive land in central cities. Finally, and critically, his ideas of land expropriation and cooperative ownership challenged bedrock American dogmas of private property. Henry Wright said of Le Corbusier's vision that "it would require a revolution in our ideas of city building and land ownership." Even for those disposed to allow public housing, it provoked suspicions of a government takeover of a sacrosanct reserve of the private sector. For the majority perhaps, public housing was socialistic, or worse, totalitarian.

In other words, America, his ultimate inspiration, was not ready for Le Corbusier.

. . . .

Ever so slowly, the wall of opposition to public housing was eroding. The government had first ventured into building housing for munitions workers toward the end of World War I. In the Depression, President Franklin Roosevelt's 1937 Housing Act offered cities financing for slum clearance, with the provision of one-to-one rehousing. Meant as an employment program, not a housing program, it had only modest impact, mostly in cities like New York willing to demolish slum areas and replace them with new housing, though it was still low-rise, made of brick, and small-scale. With new federal money available, large-scale slum clearance caught the eye of some downtown interests. A harbinger was seen in Miami, where white and black leaders alike coalesced around a plan to tear down 340 acres of Colored Town (later called Overtown) northwest of downtown, home to most of Miami's 25,000 black citizens. The evicted residents were relocated five to six miles away on the city's outskirts, in the 243-unit Liberty Square project, the first public housing built for blacks. For white property owners it was a steal: American taxpayers footed the bill to remove undesirable people from downtown, boost property values, and open valuable real estate for development. For the black community, it was the beginning of decades of wholesale removal from the city center to the suburban fringe, as if it were a defeated Indian tribe marched from its home onto a reservation.

In cities nationwide, anxiety about suburbanization had begun to replace confidence in the inescapable dominance of the central business district. People weren't coming downtown in the same numbers, but shopping, banking, and seeing movies at regional suburban commercial centers. Downtown property values and retail sales were declining, while vacancy rates rose. Commercial property owners saw themselves increasingly surrounded by residential districts growing poorer and darker skinned—people not their customers, and alarming to their customers. What was needed was to somehow attract white people, especially white women, back from the suburbs. Doing so would require offering them a bit of suburbia to go with the advantages of proximity to the center. Le Corbusier's towers-in-the-park were just right: superblocks cut off from the street grid, providing safe, green oases from the city streets and their denizens, within walking distance of the attractions of downtown.

Gradually, the capitalists—at least those downtown—rallied to the Corbusian vision. Tall concrete slab construction was already common for commercial and industrial buildings, and making inroads in housing, even in the private sector. Three Y-shaped apartment towers with traditional detailing at Alden Park, built during 1925–28 on the fringes of Germantown, Pennsylvania, pointed the way. Working for the NYCHA, Swiss-born William Lescaze and his assistant, Albert Frey, another Swiss who had worked in Le Corbusier's atelier in Paris, had prepared Corbusian designs for public housing in the early 1930s, and Lescaze is credited with the Williamsburg Houses in Brooklyn (1934–37), 20 4-story brick tower blocks with clear International Style influence. Four years later, the NYCHA completed the East River Houses in Harlem: 6-, 10-, and 11-story towers in a superblock arrangement, the first towers-in-the park project in New York City, but far from the last.

European architects were early importers of modernism. In 1928, an Austrian transplant to Los Angeles, Richard Neutra, exhibited his Rush City Reformed, a Radiant City on steroids, with ranks of identical slabs lined up to the horizon—nine years before the Bauhaus modernists Ludwig Mies van der Rohe, Walter Gropius, and

Marcel Breuer arrived in America bearing the gospel. The CIAM principles of functional segregation, carried by a growing, and increasingly influential cadre of European architects fanning out across the world, including Gropius, Alvar Aalto, and Josep Lluís Sert, gradually gained currency among city planners and others. By the late 1930s, the machine aesthetic had captured the popular imagination. The 1939 New York World's Fair gave 44 million visitors a peek at "the world of tomorrow," which included industrial designer Norman Bel Geddes's Futurama exhibit: 36,000 square feet of a planned worldly Utopia funded by General Motors, complete with automated freeways linking bucolic rural landscapes with suburbs and a decidedly Corbusian "city of the future" of skyscrapers in parks, stacked expressways, and elevated pedestrian sidewalks. "Residential, commercial, and industrial areas have been separated for greater efficiency and greater convenience," observed the accompanying promo film. This "wonder world" was "the American scheme of living."

The vision was captivating. The problem was its cost. The "outmoded commercial areas and undesirable slums" the film proposed to replace remained persistently expensive, a combination of good returns from slum rents and landlords' expectations of sale to an expanding central business district—far too costly for private developers to absorb, even if they could assemble enough contiguous lots to make a superblock. With land plentiful and cheap on the periphery, they had no incentive to build in central cities. The answer was equally clear: government would have to buy the land at inflated market values and sell to private developers at a discount low enough for them to make a profit. The economics would not support building housing for the poor. Indeed, the subsidies rarely targeted the worst slums in a city, which were often far from downtown, and in some cases targeted areas that weren't slums at all. It would mean evicting thousands, or tens of thousands, of working-class and poor residents to make way for housing for the middle and upper classes—with billions in federal subsidies—all to prop up property values in the central business district.

It was a remarkable idea, but a hard one to sell. Its boosters had to invent a new concept: blight. Blight described an area that was not—or not yet—a slum, but beyond that the diagnosis was slippery. Some defined blight as an area where the buildings were old, or perhaps dilapidated. Even this was not a requirement: it could be defined simply as an area where property values were declining, stagnant, or even rising, but more slowly than in other areas. One Philadelphia planner defined blight as "a district which is not what it should be." But increasing numbers of Americans agreed that a "blighted district" was a menace: a potential slum, and therefore a threat to public morals, health, and even the city's existence. Blight was routinely labeled a "civic cancer," deemed "infectious," a diseased part of the city that must be cut out with "the surgeon's knife" before it could spread. A Philadelphia judge opined: "If the cities are to live they must remove the blighted areas, which like a cancerous growth would eventually destroy them." Beyond that, blight was a threat to fiscal integrity: blighted areas used more in services than they paid in taxes, three to five times as much, some commentators warned. Replacing blighted districts with wealthier residents in new apartment buildings would see property taxes, hitherto disappearing to the suburbs, flow back into municipal coffers.

The idea gained traction as a set of disparate forces began to align: housing reformers, downtown real estate interests, unions, the building industry, city planners and academics, big-city officials, even some advocates for the poor, who saw the need for dramatic intervention.

The prominent public housing advocate Catherine Bauer wrote: "If we are to build houses and cities adequate to the needs of the 20th century, we must start all over again from the ground up." She and others insisted that the purpose of any clearance be rehousing, but the central business district interests, and most Americans, were perfectly willing to clear slums not to rehouse the poor—which they considered socialism—but to redevelop and expand downtown for the better-off. A series of state laws were passed, the first in New York in 1942: the Redevelopment Companies Law, enabling government redevelopment agencies to use the power of eminent domain to assemble blocks of land by buyout and eviction, then turn the land over to private developers—at a discount sufficient for them to build for-profit housing. It was socialism for the well-to-do.

The first major project was Stuyvesant Town, built by the Metropolitan Life Insurance Company, the largest corporation in America at the time, drawn to the housing business because its directors believed that well-housed people would live longer, thereby improving the company's bottom line. One writer called its development model "the business welfare state." The target was 18 square blocks in the Gas House District, a mixed industrial-residential section of the Lower East Side, once teeming with immigrants, which had lost half of its population by the 1920s. New York City's powerful public works czar Robert Moses, who had spearheaded the building of the Long Island parkway system, Jones Beach, and the Triborough Bridge during the Depression, exercised eminent domain, transferred the land at discount, and gave the company the streets, along with a 25-year tax freeze. Metropolitan's Utopia was Corbusian: the neighborhood was leveled—observers said it looked as though it had been bombed—its 10,000 inhabitants displaced (most far away, as few found affordable rentals nearby); then the area was made into a single, 60-acre superblock stretching from Fourteenth to Twentieth streets and First Avenue to Avenue C. When finished, 24,000 people were housed in 35 slab buildings set amidst lawns, trees, concrete pathways, and parking lots—a "suburb in the city." Metropolitan Life barred nonwhites from residing there, racial segregation being a central attraction of suburbia in the first place. A newspaper headline blared: "Giant Housing Project Which Wiped Out Slum a Masterpiece of Capital."

A few other similar efforts were made. Pittsburgh Equitable Life Assurance Society built Gateway Center on 23 acres downtown. Besides insurers, developers and banks weren't eager to invest. A federal program was required, and it arrived with the Housing Act of 1949, signed by President Truman that summer. The meat of the act was its Title I, which not only pegged the government's share of the cost subsidy at two-thirds, but allowed developers to build anything they wanted, if it could be called the land's "highest and best use." That might mean convention centers, office buildings, stadiums, or luxury apartments. Often it meant parking lots—because, besides the poor and working-class living nearby, downtown's biggest problem was cars.

By the early 1930s, most American central business districts were saturated with cars, causing traffic gridlock and parking nightmares so severe that suburbanites simply "stopped going downtown" rather than take public transport. Downtown interests believed they could solve the problem by building freeways: new limited-access highways, radiating outward from the downtown hub, to lure people in with fast transportation. Critics scoffed, pointing out the dubious wisdom of building often-elevated freeways that carried 6,000 people per hour while cities were tearing out elevated railroads that carried 40,000 people per hour. Many predicted that ramming

300-foot-wide freeways through neighborhoods would displace thousands and make already declining areas even worse. The effort might cost $3 million per mile, or hundreds of millions to rework even midsize cities. But the downtown coalition pushed forward: planners drew plans, and officials successfully drew in state gas taxes and federal grants. Oakland got the Nimitz Freeway, Boston the Central Artery; New York City got the elevated West Side Highway and the Gowanus Parkway, which Robert Moses realized by tearing down the old Third Avenue El train through the working-class Sunset Park neighborhood of Brooklyn, replacing it with four traffic lanes elevated over ten lanes below, destroying 100 businesses and moving 1,300 families. The neighborhood never recovered.

Nationwide, the result was a three-pronged assault on the cities, broadly falling under the domain of "Urban Renewal" (though the term wasn't coined until 1954): Title I urban redevelopment with government-funded clearance and private rebuilding; public housing construction, under the Title I program and separately; and a massive highway building campaign, connecting cities with multilane freeways and, after lobbying by the cities, also penetrating them with radial arterials from downtown to the suburbs and encircling them with beltways. The effort accelerated with passage of the 1956 Federal Highway Act, signed by President Eisenhower and raising the federal cost share to an irresistible nine-tenths. With such enticements, few cities resisted, and Urban Renewal, unwinding over decades, radically changed the fabric, character, demography, economy, and politics of the American urban landscape.

New York City was perhaps the prime example: under Robert Moses, the celebrated and reviled "master builder" who headed a series of public authorities during his 1934–68 reign, the metropolitan region was rebuilt to serve the interests of automobiles, suburban commuters, and the central business district. Though unelected, Moses controlled toll revenues and could issue bonds based on them, putting his agenda beyond legislative or public involvement. Beginning as commissioner of public parks, he built parks, swimming pools, and public beaches to serve hundreds of thousands of people (though black people were excluded, as was customary at the time). An early admirer of Le Corbusier, he eventually became one of his most fervent and effective disciples; declaring that "cities are created by and for traffic," he dedicated his career to realizing his mentor's vision of towers-in-the-park served by a regional network of fast roads. He built bridges over the Hudson, the East River, and the Harlem River, and more highways linking Manhattan to New York's other boroughs and its suburbs. Believing that the new Interstate Highway System "must go right through cities and not around them," Moses planned to build yet more through its heart. And he cleared blighted districts: 9,000 acres of them, comprising 17 Title I projects and a host of others under separate programs—earning the sobriquet of "the king of Title I." Infamously, he said, "You can draw any kind of picture you like on a clean slate and indulge your every whim in the wilderness in laying out a New Delhi, Canberra or Brasilia, but when you operate in an overbuilt metropolis, you have to hack your way with a meat ax."

Much of Moses's hacking benefited the elite. Even with a federal "write-down" of two-thirds of the land's cost, New York's expensive real estate dictated that most housing built would be high-end. Exclusive universities founded new campuses where homes had been: New York University moved from the Bronx to Greenwich Village, where it took over the south side of Washington Square; Long Island University, the Pratt Institute, Fordham University, and the Juilliard School also received presents from Moses. On the

West Side, the site for Lincoln Center was cleared, yielding opera, dance, and luxury high-rises, at the cost of the homes of thousands of lower-income New Yorkers. In all, Moses moved 200,000 people to public housing, typically far from their old neighborhoods. One study following the first 500 evicted families noted that 70 percent moved outside the neighborhood, the average rent rose 25 percent, and just 11.4 percent were moved into public housing. In Manhattan alone, his programs cleared 314 acres, built 28,400 market-rate apartments, and 30,680 public housing units in Title I projects and 21 others adjacent to them. The new projects were modernist superblocks, with closed cross streets and the surrounding avenues widened to absorb the rerouted traffic, making the project's separation from the city even worse. Where the old land coverage had been 80–90 percent, Urban Renewal's tended to be about 30 percent, much of the "open" space being covered with surface parking and sometimes underground garages.

For those communities that were "redeveloped" or "renewed," Urban Renewal was devastating. Despite its stated purpose in the 1949 Housing Act, clearance tended to radically reduce housing stocks and drive up prices for what remained. On average, just one unit was built for every four units demolished. The election of President Eisenhower in 1952 lowered the priority of the program's housing component even further, as the general was more interested in building interstate highways. Passage of the 1954 Housing Act gutted the housing provision, so that in many cases, people were evicted with nowhere to go. The impact was felt disproportionately by African-Americans: Urban Renewal was really "negro removal," jibed the writer James Baldwin. Indeed, entire communities were erased, like the Africville neighborhood of Halifax, Nova Scotia, cleared during 1964–70 to make way for a highway interchange and a bridge. Even where parts of neighborhoods were spared, highway construction often bisected them, cutting residents off from other parts of the city, effectively isolating them physically and economically, as low-wage industrial jobs disappeared or followed fleeing white residents and businesses to the suburban fringes. The result was a downward spiral of falling property values and rising unemployment and crime. New Orleans's troubled Treme district, cut off by the construction of Interstate 10, is typical. The 1965 Watts riots in Los Angeles, which engulfed 46 square miles of the city over six days, leaving 34 dead, followed the isolation of African-American areas by I-110 (completed in 1962), I-10, and I-710 (both completed in 1965). New York City's Bronx, a vibrant borough of mixed residential neighborhoods, commercial and industrial areas, began its decline after Robert Moses hacked the Cross Bronx Expressway through it, from 1948 to 1972 (one of many highways built through the borough), the borough would soon become an internationally recognized symbol of urban failure.

Many of those who were left were relocated to gigantic public housing projects, stranded in superblocks from the rest of the city, often surrounded by highways and parking lots. In Saint Louis, Minoru Yamasaki designed the Pruitt-Igoe project, consisting of 33 11-story slabs (1950–56). Detroit got the Gratiot project; Cleveland the Hough; Chicago its Cabrini-Green and Robert Taylor Homes—just to name a handful of the biggest and most notorious. With some exceptions, these projects quickly deteriorated, through a combination of inadequate upkeep and policing by cities cutting services as their tax bases moved beyond municipal boundaries to the suburbs, spiking unemployment and the family breakdown that attends it, and crime in common areas such as corridors and stairwells, and especially the deserted open spaces between buildings, too large and impersonal to be kept secure. Better-off residents left, exacerbating the problems. By the 1970s, racial segregation, arson, crime, and abandonment came to epitomize the landscapes of many American inner cities—outside of the well-policed new high-rise business districts. Places like the Bronx, where landlords burned down their tenements by the thousands for the insurance payouts and hundreds of blocks lay abandoned, were routinely compared with bombed-out war zones. Urban Renewal, specifically Le Corbusier's design prescription for it, had proven a very costly failure. The most notorious projects were torn down, most spectacularly Saint Louis's Pruitt-Igoe in 1972–73, and Chicago's Robert Taylor Homes in 1998.

Urban Renewal by itself didn't cause the decline of central cities. The movement away from city centers had been under way for more than a century, while deindustrialization was a global phenomenon. A complex web of public policies subsidized and accelerated them, seeking to make the lives of some parts of the population better, while making those of millions of others, mostly nonwhites, far worse. And yet architecture and urban planning must share a large portion of the blame, adhering as they did blindly to the dogmas of prophets of the future Utopia like Le Corbusier. His design tenets—the separation, exclusion, segregation, of kinds of activities and kinds of people—led directly to the results achieved. A typical example is Downtown Los Angeles's Bunker Hill redevelopment project, which replaced an older, "blighted" residential area with a superblock high-rise business district surrounded by concrete palisades and freeway interchanges. Le Corbusier did after all want to end the city as it existed. His city of tomorrow was indeed, as Marshall Berman and others have written, urbicide.

3. Slabs

Field Guide: Tower Blocks

Diagnostics:

- Slabs are tall buildings, usually modernist with minimal detailing, repetitive facades and windows, and straight sides. Usually glass walled or of cast concrete. Residential, office, or institutional uses, but usually a single use per building.
- They are set apart from other buildings, generally surrounded by nonbuilt space, whether open or park space, parking lots, or roads.
- They are connected by roadways to fast arterial roads, viaducts, or highways. Car-dependent.
- City plans are regular, monumental arrangements of slab buildings and fast roadways, viaducts, or highways.

Examples:

USA

- Boston: Government Center.
- New York: United Nations Secretariat (1947), Lever House (1952), Seagram Building (1958), Stuyvesant Town, Lefrak City, New York City Housing Authority projects.
- Pittsburgh: Golden Triangle redevelopment.
- Cleveland: Riverview, Erieview.
- Los Angeles: Park La Brea, Bunker Hill redevelopment, Century City.

United Kingdom

- London: Housing estates: Alton Estate, Roehampton; Peckham Estate, Southwark; Elgin Estate, North Paddington, etc. Office blocks: London Wall, Centre Point, Euston Center, Stag Place, Camden Towers, etc.
- Glasgow: Red Road Flats (demolished).
- Newcastle: Byker Wall.

France

- Paris: La Défense, Montparnasse Tower, Olympiades complex, etc.
- Nationwide: HLM, habitation à loyer modéré (moderate-cost housing).

Including nine Villes nouvelles (New Towns): Cergy-Pontoise, Marne-la-Vallée, Sénart, Évry, and Saint-Quentin-en-Yvelines on the outskirts of Paris, and others outside Lille, Lyon, Marseilles, Rouen, and Grenoble.

Russia and the Former Soviet Union

- Moscow: Kalinin Prospekt, Novy Arbat. Slabs are ubiquitous in Russia and all former Soviet republics, from Tallinn, Estonia, to Tashkent, Uzbekistan; Bishkek, Kyrgyzstan, to Tbilisi, Georgia; Almaty, Kazakhstan, to Yerevan, Armenia.

Eastern Europe

In the Soviet-aligned countries of the Warsaw Pact and other client states, the same idiom prevailed, becoming the ubiquitous urban form.

- Germany (eastern part, former GDR): The ubiquitous slab apartment blocks are called Plattenbauten, "panel buildings."
- Hungary: Here the slabs form what is called a Panelház and house 2 million people, a fifth of the country's population.
- Poland: Simple slabs were extended into curving lines called Falowce, or "waves," because of their shape. One such building in the city of Gdańsk is 11 stories tall and nearly 3,000 feet long, and shelters 6,000 people.

Images:

"Corbusierhaus," Unité d'Habitation (1957), Berlin, Germany. Architect: Le Corbusier.

United Nations Secretariat and General Assembly (1952), New York, New York. Architects: Le Corbusier, Oscar Niemeyer, and others. *Norbert Nagel / Wikimedia Commons; License: CC BY-SA 3.0*

Stuyvesant Town and Peter Cooper Village (1947), New York, New York.

Stuyvesant Town (1947), New York, New York.

Pruitt-Igoe Houses (1955), Saint Louis, Missouri.

Pruitt-Igoe Houses, Saint Louis, Missouri. Demolished 1972.

Park Hill housing estate (1957–61), Sheffield, South Yorkshire, England.

Falowiec, "wavy building" (1970s), Gdańsk, Poland. *© 2005 by Johan von Nameh*

Panelák, "panel building," Prague, Czechoslovakia. More than a million panelák flats were built between 1959 and 1995, housing one-third of what is now the Czech Republic's population.

Kin Ming housing estate (2003), Hong Kong. Housing 22,000 people.
Author: Baycrest-Wikipedia user/ Baycrest–維基百科用戶; license: CC-BY-SA-2.5

Housing projects (1970s–1980s), Ulan Bator, Mongolia. Called "Ugsarmal" in Mongolia, built using Soviet designs.

Slab housing (view 2004), Shanghai, China.

Chapter [4]

Homesteads

Frank Lloyd Wright and the Anticity

Utopia has long been another name for the unreal and the impossible. We have set utopia over against the world. As a matter of fact, it is our utopias that make the world tolerable to us: the cities and mansions that people dream of are those in which they finally live.

—Lewis Mumford, *The Story of Utopias*

At 8:30 p.m. on Monday, April 15, 1935, in the Oval Room of the White House, President Franklin Delano Roosevelt pressed his finger down on a gold telegraph key to launch a cascade of electric pulses north to New York City. Seconds later, 120 flashbulbs popped loudly in a room in Rockefeller Center on Fifth Avenue, 50 floodlights lit up, a siren wailed, and an electric organ began to play while an American flag dropped down from the ceiling. Thus was marked the official opening of the Industrial Arts Exposition of the National Alliance of Art and Industry, a show of mass-produced consumer goods chosen specially for display because their designs "emphasized beauty." The nation was deep in the Great Depression, and the exposition aimed to revive hope and economic activity by showcasing the leading edge of American manufacturing and innovation. Driving it were stunning, seemingly magical advances in harnessing electricity—in appliances, tools, telephones, and radio. A crop of massive new buildings, decorated with Art Deco motifs that mixed Egyptian and classical patterns with modern streamlining, was rising to symbolize and encapsulate the country's technological resurgence. In New York, to join the Empire State Building was Rockefeller Center, including Radio City Music Hall, finished in 1932, with other parts of the complex still under construction in 1935. Its western counterpart was Boulder Dam, a smooth concrete arch looming 700 feet over the Colorado River in the forbidding Nevada Desert, the largest dam ever built, which would be dedicated on September 30 of that year. Electricity from its generators would cross 250 miles of desert and mountains to light up the night sky over Los Angeles.

Earlier that Monday, the country's most celebrated architect, Frank Lloyd Wright, gave a speech recorded by a microphone inside Rockefeller Center for broadcast on radio. Wright had become famous decades before, both for his association with his former employers and mentors, the Chicago skyscraper builders Louis Sullivan and Daniel Burnham, and for developing his own groundbreaking Prairie Style houses, identified as a uniquely American form based on innovative uses of concrete and glass. Wright's stature had once rested on his architecture's bold modernism, but his best work was behind him, done in the 1890s and 1900s, before he was 40. Aged 67 in 1935, he was best known by the general public for the scandals that had engulfed him since 1909, when he abandoned his wife and children and ran away to Europe with the wife of a client, followed by a string of disasters, including the murder of his family by a mad servant, fires repeatedly destroying his home, jail time, divorce, and bankruptcy—well-publicized crises more captivating than even the most futuristic buildings.

Wright cut a figure commensurate with his public image: wearing capes, hats, balloon ties, and carrying a cane ("entirely superfluous" in the words of his biographer Meryle Secrest), he was reliably ostentatious, irascible, and provocatively arrogant, sure in his creative superiority and vision for the world. His speech didn't disappoint. Calling out Rockefeller Center as a concretion of despised corporate power—"the entrails of final enormity" as he put it—he began not with the subject of design but with his critique of America's economic and governing structure. "I love America and her idea of democracy," he began, but the country had descended into "economic, aesthetic and moral chaos," its people herded into gigantic, squalid cities and dehumanized by a behemoth industrial capitalism. America's leaders had failed to meet the challenge of the machine, putting the basis of American democracy—the freedom of the individual—at risk. "Meantime, what hope of democracy left to us goes from bad to worse," he railed. "Here we are . . . tragic breakdown staring us in the face. The present success-ideal proves to be a bad one for all but the few."

Wright then announced that he, by virtue of his skill as an architect, could envision another way: "I have tried to grasp and concretely interpret the whole drift of great change taking place in and around us in order to help create a human state more natural than the one that present cupidity and stupidity will allow." That more natural state he called Broadacre City, the "great metropolis of the future," which would replace not just existing metropolitan monsters like New York City, but smaller cities and towns as well, putting in their place a decentralized, unbounded landscape that blended the best features of both. In his characteristically cryptic fashion he described it thus: "The city will be nowhere, yet everywhere."

There in the exposition hall was a model of Broadacre City: 12 feet by 12 feet, 8 inches, at a scale of 1 inch representing 75 feet, equaling an area of 2,560 acres, or four square miles. Laboriously constructed of cardboard and wood by the 25 apprentices in his Taliesin Fellowship in Wisconsin and Arizona and then hauled to New York in a truck, it was intricate and precisely detailed, its surface accentuating shadows and contours, like a bas-relief. Broadacre City didn't look like a city, but like bits of a small city center and a small town scattered about in a mostly flat countryside, the widely spaced buildings almost lost in vineyards, strips of orchard, blocks of woods, and farm fields. Spread across the agrarian landscape was a modified grid containing distinct areas of family homesteads, small manufacturing buildings and laboratories, and larger buildings housing schools, a church, a small university, and a hospital. Alongside fields and lakes were small, isolated apartment towers. Elsewhere were public facilities: zoo, aquarium, arboretum, airport, arts building, county seat, county fairgrounds, and sports fields fit into the gentle contours of the land. Along the length of one side of the model ran a multilevel, multimodal transportation corridor called "the great arterial," stacked and segregated according to use, like Le Corbusier's seven-level transport corridor: a two-tier highway with 10 car lanes above and 2 truck lanes below, in the median above which ran a high-speed monorail. The arterial followed the path of, and would in most areas take the place of, existing freight railroad tracks, little used since coal would be burned for electricity at the mine head—instead of the city—and transmitted over new long-distance wires, like those extending across the desert from Boulder Dam. Below the highway was continuous warehousing, and alongside it was strip development of markets, industry, and motels. Stretching away on the orderly grid were smaller roads, gradually yielding to country roads. Broadacre City depended on personal cars—every family would have one—and in later versions of the plan a smaller number of "aerotors," futuristic-looking personal helicopters.

Wright was at pains to explain that "Broadacres" wasn't an alternative to the city, but a complete replacement for it: A NEW PATTERN OF LIVING FOR AMERICA, according to a slogan plastered across a display panel. "Broadacre City is no mere back-to-the-land idea but is, rather, a breaking down of the artificial divisions set up between urban and rural life." It would do away entirely with the notion of an urban or even town center, scattering the center's traditional functions around the landscape. Occupying the middle of the model square was the grid of homesteads. By "homestead" Wright meant literally a complete, integrated agrarian enterprise comprising a single-family residence, a small farm formed of up to one acre per person in the household, and perhaps a small factory or laboratory. Each head of household would be a combination of farmer and mechanic, scientist, or intellectual. The houses were to be assembled from prefabricated components, by the owners, customized to suit their needs. Wright assumed there would be some income inequality in his Utopia, so he provided for "minimum, medium, and larger" houses, each with different sizes of garage; thus, as he put it, "1-, 2-, or 5-car" houses. For those not pursuing farming, the apartment towers would house workers laboring in larger industrial enterprises. But the solid majority of Broadacres' citizens were to be single-family homeowners, by design. "The true center (the only centralization allowable) in Usonian democracy is the individual in his true Usonian home," Wright wrote in 1945, referring to a later version of Broadacres. ("Usonia" was his coinage for his redesigned, better United States—a phrase he may have borrowed from the utopian writer Samuel Butler, or simply made up.) As early as 1910 he had asserted the single-family home's primacy: every American had "the peculiar, inalienable right to live his own life in his own house in his own way." In short, not only was a man's home his castle, but the single-family home was "the only permissible shelter for a free society," in the words of historian Robert Fishman.

Broadacre City wasn't restricted to the four-square-mile community depicted on the model, which was meant only as a generalized reference. It would blanket the nation, limited only by topography and water, a continuous expanse of "villages," each housing 1,400 families and organized as an autonomous county, each village-county roughly 20 miles apart along "the great arterial," which, Wright assured, would be "great architecture." "Broadacre City is not merely the only democratic city," he wrote. "It is the only possible city, looking forward to the future." Wright's diorama was a diagram in plan form of his belief in radical decentralization as the solution to America's problems, a belief he shared with a wide, diverse swath of his countrymen, imbued with his own libertarian convictions. "In Broadacres you will find not only a pattern for natural freedom for the individual as individual. You will find there structures based upon decentralization of nearly everything big business has built up to be big, and you will find an economic ground-structure aimed at more individuality and greater simplicity and at more direct responsibility of government." Government would be "reduced to one minor government for each county," taking the form of a county architect, responsible for managing the distribution of land (all owned by the state), providing services, and guaranteeing the high standards of design and construction that Wright promised as a necessary bulwark against "economic, aesthetic and moral chaos."

On another display board posted beside the model, Wright provided a list of some fundamental design conditions, as if enshrined in a constitution, along with various directives, promises, and diktats. All began with "No." It was affirmation by means of negation:

No private ownership of public needs
No landlord or tenant
No "housing." No subsistence homesteads
No traffic problems
No railroads. No streetcars
No grade crossings
No poles. No wires in site
No headlights. No light fixtures
No glaring cement roads or walks
No tall buildings except in isolated parks
No slum. No scum
No major or minor axis

What did it add up to? Clearly, Broadacre City was part of the long tradition of utopian visions of a rural Eden: bucolic, low density, agrarian, Jeffersonian. Wright was determined to build smallness into its DNA, with "little farms, little homes for industry, little factories, little schools, a little university going to the people." The plan countered the concentrating and monopolizing tendencies of large scales, whether urban or industrial. Each county would be largely self-sufficient, with most things produced locally and consumed locally: "The waste motion, back and forth haul, that today makes so much idle business is gone." As a solution to the economic depression, putting people to work to satisfy their own needs made intuitive sense to Wright.

In its attention to arcadian aesthetics, Broadacre City was a direct descendant of the romantic garden suburbs of Frederick Law

Olmsted, with which Wright was well familiar, certainly from Riverside, Illinois, near his home in Chicago. In its combination of agriculture, industry, residences, and greenbelts as a kind of cordon sanitaire against bigness, the whole serviced by a modern transportation matrix, it was like Ebenezer Howard's garden city—itself influenced by Howard's time in America, including on a Nebraska homestead, where he would have soaked up both the romantic garden suburb and the idealism of homesteading the frontier, before bringing them back to soot-choked, industrial Britain. But Broadacre City had no conventional center. Transport, industry, and institutions were shunted to the sides. There was no Crystal Palace to focus commercial and community life. Instead, shopping would take place at the "roadside market" where people would sell their own products directly as well as buy from others, eliminating the middleman as the county architect eliminated the rest of government. Wright provided for a "community center," which would contain restaurants, an art gallery, theaters, a golf course, a racetrack, a zoo, an aquarium, and a planetarium—an "attractive automobile objective," he wrote, and what would later be christened an "experience" shopping mall.

Wright's conception went beyond other earlier visions of the rural Eden in integrating modern technology; indeed, Broadacre City was predicated on it: "By a more intelligent use of our developed scientific powers we establish a practical way of life that will bring the arts, agriculture, and industry into a harmonious whole." Wright listed three key new technologies: "The three major inventions. . . . 1. the motor car . . . 2. radio, telephone, and telegraph . . . 3. standardized machine shop production," as spelling the end of the concentrated city, "whether the powers that be that over-built the old cities like it or not." Like Le Corbusier, he was convinced that these powers had rendered the city "no longer modern," in his words. But, unlike the Swiss modernist, Wright's embrace of the machine served decentralization, not concentration. Throughout his career, he excoriated Le Corbusier and the other European modernist architects as agents of centralizing, undemocratic authority. In contrast, his lifelong project was searching for a means of coexistence between his cherished ideal of Jeffersonian, quasi-agrarian democracy and industrial methods. Since at least his 1901 lecture "The Art and Craft of the Machine," delivered at Hull House in Chicago (Jane Addams's settlement house and a nerve center of the broad reform coalition aligned with the Arts and Crafts movement), Wright had sought a way to harness industrial technology to the movement's project of reconnecting work and life lived in harmony with the land—in essence, trying to bring the machine safely into the garden. He was very aware of the damage these technologies had already wrought: "The price of the major three to America has been the exploitation we see everywhere around us in waste and ugly scaffolding that may not be thrown away." But Wright assured his audience that his system of a county architect—a clear surrogate of himself—would guarantee "good architecture" and, through it, good policy.

Of all the inventions, it was the automobile that most possessed him. Its speed—60-mile-per-hour cars versus Howard's walking or bicycling pace—gave mankind "MASTERY OVER TIME AND SPACE," and unshackled the city from its traditional limits. "The door of the cage is opening as one consequence of the motorcar invasion and of collateral invention," he would write in his 1945 book *When Democracy Builds*. "The actual horizon of the individual immeasurably widens. And it is significant that not only have space values entirely changed to time values with the new standard of measurement but that the new sense of spacing is truly based upon

Mobility. Mobility is now at work upon the Man himself in spite of himself. And, too the impact to the modern sense of space it engenders is spiritual as well as physical." The car, to Frank Lloyd Wright, a man who had spent decades trying to outrun his constraints, whether wives, creditors, conventional morality, or the wrong aesthetics, granted liberation. "If he has the means to go, he goes. He has the means in his car."

In the carefully sawn and glued model of Broadacre City lay a nesting set of contradictions. Wright's Utopia promised to tame the destructive force of the modern industrial metropolis by yoking the machine to a model nineteenth-century rural agrarianism, thereby achieving a more perfect modernity. It would redeem the city by dismantling it and prohibiting the conditions for it to form again. It would protect society and defend democracy by privileging the nuclear family, and mostly, the free individual—but do so through the total, if enlightened, despotism of one unelected person.

As impressive as Wright's vision was, what it demonstrated most clearly was his failure, or at least marginalization, as an architect—the Broadacre City model was virtually the only thing Wright built in the first half of the 1930s. After the enormous success of his earlier career, marked by the triumphs of the Prairie houses (1890s–1909); the Larkin Administration Building (Buffalo, New York, 1903), which helped pioneer the use of air-conditioning, plate glass doors, and metal furniture; Unity Temple (Oak Park, Illinois, 1908), an early experiment in poured concrete; and Midway Gardens (Chicago, 1913), his commissions began to drop off, owing in part to the notoriety of his private life, and in part to architectural fashions shifting in other, more traditional directions. In his peregrinations after the scandal of 1909, Wright realized several masterpieces, including four unique houses in Los Angeles built of concrete "textile" blocks (1919–23), and the Imperial Hotel in Tokyo (1923), a feat of engineering that survived that year's earthquake that leveled most of the city. In 1924, he designed an innovative skyscraper for the Chicago National Life Insurance Company to be built of reinforced concrete, sheet metal, and glass, though it remained on his drawing board. Between 1924 and 1926 he built nothing. A visitor to him in those difficult years as Wright approached 60 described him as "passive and isolated," ignored while younger architects pulled in the commissions. As if to throw earth onto his coffin, in 1932, while including Wright with the European modernists in the International Style exhibition he curated at New York's Museum of Modern Art, the firebrand young architect Philip Johnson damned him by praising him as "the greatest architect of the 19th century."

As in fables, the great man's fall began when he was most secure in his powers and fame. At the height of his career designing houses for the well-to-do bourgeoisie in Oak Park, a leafy suburb of Chicago, Wright was the toast of the town. He fairly embodied the values of comfortable, prosperous domesticity that he designed into his clients' homes, beginning with the magnificent hearth placed at the center of each one. Living in his own, self-designed house with his wife, Catherine, the daughter of a wealthy client, and six children, he fit the role of a model paterfamilias. Then, in 1907, Wright fell in love with 28-year-old Mamah Borthwick Cheney, the wife of another client, Edwin Cheney, with whom she had two children. The two were seen cavorting around town in Wright's car. Catherine, somewhat inured to his wandering, refused to give him a divorce. In 1909, he assigned his projects to an associate in his office and left with Mamah for Europe, in the process dropping an enormous commission from Henry Ford, surely the job of a lifetime. The publicity was swift and harsh. After an eventual divorce, Wright installed Mamah Cheney and her children in a new house he built near his childhood home, in a lovely farming valley near the village of Spring Green, Wisconsin. Completed in 1911, he dubbed it "Taliesin" after a Welsh poet beloved by his mother, who had instilled in him a love of art and architecture, medieval culture, and a fierce family identification, which Wright carried on by adopting the Welsh family motto "Truth against the world" as his credo. In August 1914, while Europe marched to war, domestic horror struck at Taliesin while Wright was away, when a crazed manservant murdered Mamah, two of her children, and five others, before setting fire to the buildings. In despair, Wright set off on what would be six years largely lived on the road, building his California and Japan projects. Despite the favorable press around them and the Imperial Hotel's surviving the 1923 earthquake, he struggled to secure work. His clients, after all, had been stolid Midwestern businessmen seeking safe suburban homes, and the damage to his reputation was too much to overcome.

In 1923, he met and married Miriam Noel, but the union didn't last the year. In 1924, the partly rebuilt Taliesin was again damaged by fire. Wright was forced to mortgage it to cover his mounting debts. In November of that year, he met Olga Ivanovna Milan Lazovich, a native of Montenegro who had wandered Europe in the turbulent years before World War I and, in a failed marriage and with a daughter, then-seven-year-old Svetlana, moved to New York. Ogilvanna, as she was called, was elegant, beautiful, and cultured, with a mysterious air; from 1915 to 1922 she had followed the mystic Georges Gurdjieff across war-torn and chaotic Europe, becoming a central member of his circle. When she met Wright, both were seeking divorces. The architect was smitten. He was reported to have said to her: "Come with me, Ogilvanna, and they will not see us for the dust!" But Miriam, the second wife, also refused to grant him a divorce, and became vindictive, harassing the couple. In February 1925, Wright moved Ogilvanna and Svetlana into Taliesin. A daughter, Iovanna, was born. Miriam, angry, succeeded in having them evicted, and, in September 1926, after Wright and Ogilvanna fled to Minnesota, had them tracked down and jailed for violation of the Mann Act, alleging criminal adultery and transportation across state lines "for immoral purposes." The press had a field day: the *New York Times* called Ogilvanna a "Montenegrin danseuse" and Taliesin a "Love Bungalow." Miriam relented, and they were released, but the great architect was destitute, virtually in hiding.

Gradually, the tide began to turn. In 1927, Miriam acceded to a preliminary divorce and after a year's waiting period, Wright and Ogilvanna were married. Taliesin was seized by the bank, which attempted unsuccessfully to sell it at auction, but a group of Wright's still-loyal former clients and friends formed a corporation called Frank Lloyd Wright, Inc., bought the property back, and made a contract with Wright to work off the $43,000 outstanding debt. His prized collection of Japanese prints, purchased in Tokyo, was sold, then some of Taliesin's farm equipment and household goods. Wright's ambition, though, was undiminished. "Not only do I intend to be the greatest architect who has yet lived," he boasted, "but the greatest who will ever live." "After me, it will be 500 years before there is another." Ogilvanna proved his ideal partner, designing a campaign to restart his creativity and career, beginning with reframing his reputation in the media through his writing and establishing a school around him at Taliesin. In 1928, he published nine articles under the title "In the Cause of Architecture" in the magazine *Architectural Record*. Hungry for money and publicity, he sought out lecture tours, landing, among others, Princeton University's prestigious Kahn Lectures for 1930. These included meditations on cities that would form the foundation for Broadacre City and were published in 1931 as *Modern Architecture* by Princeton University Press. In 1931, he embarked on what he termed "The Show," a lecture series that took him to the Midwest, New York, and the West Coast, where he landed one precious commission for a house for a newspaper owner in Salem, Oregon. In 1932, he published two books: *An Autobiography*, which he'd started in 1927 but finished only with Ogilvanna's encouragement, and *The Disappearing City*. Both contained written "blueprints" for Broadacre City and what would become the Taliesin Fellowship. Each was part of the same, long-contemplated goal of re-creating an extended family around Wright as the wise father figure in a rural Eden of his design. *An Autobiography* begins with scenes from the Wright family farm, where he was born in 1867, and ends with a version of what would become Broadacre City. The planning was far along.

The fellowship was inaugurated at Taliesin on October 1, 1932, with 23 apprentices there to labor, not for pay, but for the privilege of working at the feet of the master. Their actual work was mostly farming, building and repairing the complex, cooking, and cleaning, with some hours devoted to drawing in the drafting room, most often copying from Wright's earlier, completed projects. There were few commissions, so little new architecture being done. In his Princeton lectures he'd called for "industrial style centers," much like the Arts and Crafts workshops he knew founded by William Morris and C. R. Ashbee in England, and closer to home, the Roycroft Shops and Press, in East Aurora, New York, founded by Elbert Green Hubbard, one of Wright's clients for the Larkin Building of

1903 and likely the source of his long hair and costume of hat, cape, and cane—"superfluous" but effective tools of self-promotion. Much of the fellowship's daily routine was organized by Ogilvanna, including music, dance, and a ritual quality to the tasks of food preparation, eating, and cleaning—techniques she had learned in her time with Gurdjieff's Institute for the Harmonious Development of Man, at the end of which she'd risen to the level of assistant instructor. While the Taliesin Fellowship shared some characteristics with the Gurdjieff circle, it wasn't a commune or collective. Visitors to the Taliesin compound near Spring Green described a far more hierarchical arrangement. Ayn Rand said it was "like a feudal establishment." A British visitor recognized it as "closer to an English manor house." The German modernist architect Ludwig Mies van der Rohe, who came to Taliesin intending to stay a few hours, but stayed the weekend, exclaimed, "Freiheit! Es ist ein Reich!" (Freedom! This is a kingdom!)

Like the fellowship, the idea of the Broadacre City plan had been gestating in Wright's mind and practice for decades. His first plans for communities or small cities were for private clients in Montana: the Bitter Root Valley Irrigation Company plan of 1909 and the Como Orchard summer colony project, both semirural town plans based on automobile circulation and without a traditional central focus. He had published an early version of the idea in a magazine article in 1916. His having traveled across the United States by car many times and lived in Los Angeles in the 1920s, where the car was already midwifing a new form of centerless urbanism organized around roads, single-family residences, and car-serving businesses, would have opened his eyes to their transformative potential. In

1932, he had considered problems of small-scale agriculture with the unbuilt "Little Farms Unit" and "Little Farms Tract" commissions for Walter Davidson, another client involved with the Larkin Building who hired Wright 25 years later to design prefabricated farm buildings. Again for Davidson, he designed a concept for Wayside Markets, a drive-up shopping mall. In a 1923 manifesto speculating on the future of cities in the aftermath of the Tokyo quake, he proposed a five-story height limit and car-driven decentralization: "Modern transportation may scatter the city, open breathing spaces in it, green it and beautify it, making it fit for a superior order of human beings." By late 1934, Wright's vision of Broadacre City was sufficiently enthralling to convince Tom Maloney of New York to give $1,000 for the model to be displayed at the Rockefeller Center show. Edgar Kaufmann, Sr., the Pittsburgh department store owner for whom Wright was designing the extraordinary residence Fallingwater that year, also wrote a $1,000 check to help the fellowship complete Broadacres. The first drawings showed a green and rural square crossed by small roads, and scattered suburban housing. The second draft was cruciform, with two large crossing road corridors and some gridded areas, with, as was written across the top, a "minimum of one acre to the family." In the final plan, finished in late 1934 and published in *Architectural Record* the next year, the "arterial" had moved to one side and the homesteads moved to the center. Work began on the Broadacre City model in the Wisconsin winter and finished at the fellowship's new workshop in the Arizona desert in the early spring of 1935. The fellows spent hundreds of hours carefully cutting cardboard and balsa, giving Wright's vision form. One fellow, Cornelia Brierly, wrote at the time: "We live in this future city. Speed in the shady lanes of its super-highway. Know the repose of its floating lake-cabins and when our backs ache and our eyes smart from bending over this finely detailed work we lose our pent up energies by romping in the grass of the courtyard."

In many ways, Wright's vision wasn't radical but mainstream—within the reforming tendencies of American culture. Certainly it was part of a long tradition of rural utopianism that traced its roots back to the Puritans and their ideal of the City on a Hill, a built environment that embodied a moral covenant, not only with God, but with nature: America was the new Garden of Eden, the sublime wilderness to be transformed by the agrarian labor. Wright's championing of it fit in with another American tradition, the jeremiad, according to the historian Narciso Menocal: "Like the prophet of old reminding the children of Israel to keep the faith among the Babylonians, American Jeremiahs preached the need to live according to an eternal and immutable ideal established by nature, one that in the end would defeat the artificiality of the traditions of humankind, on which rested the tyranny of history." Frank Lloyd Wright's maternal grandparents, Richard Jones and Mary Lloyd, had been vigorous examples of this ideal: pioneers and religious paragons arriving in the Wisconsin valley and hewing and sowing an exemplary life, like the pilgrims, or like the biblical Abraham and Sarah. To Wright, it was the "ancestral valley," even though his immigrant parents bought the land just one year before his birth. The Jeffersonian identification of the individual rural proprietor with moral superiority and political independence descended from this religious tradition and perfectly fit Wright's idea of who he was—throughout his life, he saw himself as a Jeffersonian farmer. And it lent itself to a fixation with the paradoxical idea of pastoral cities coexisting with nature in an unspoiled, productive landscape. The Transcendentalists had ruminated on it at length: in Henry David Thoreau's *Walden*, and Ralph Waldo Emerson's "Nature." At its base, Wright's philosophy rested on an idealized conception of the sovereign individual: "There is no such thing as creative except by the individual. Humanity, especially on a democratic basis, lives only by virtue of individuality. The whole endeavor, the whole effort of our education and our government, should be to discover first—then cherish, use and protect the individual."

Utopian novels were one of the most popular literary forms of the nineteenth century in the United States, beginning at least as early as James Reynolds's 1802 book *Equality*, which had all the ingredients: universal education, women's liberation, a noncompetitive economy, services provided by a central government, democratic rule by the wisest, and the adoption of labor-saving technologies. It was a century of destabilizing industrialization and urbanization, so it is no surprise that the genre of utopian romance relied heavily on depictions of benevolent machinery and cities that were really parts of the country, providing a safe and stable physical and moral environment. By century's end, the utopian novel was America's leading literary genre. This literature wanted to accomplish the reform of social relations through the restructuring of the physical world. It was to be revolution through city planning.

Wright imbibed an amplified version of it in the Progressive intellectual culture of his family: at the dinner table, at his aunts Jane and Nell's Hillside Home School nearby, and the house of his uncle Jenkin Lloyd Jones, a leading Unitarian minister, where he first met Jane Addams. The Chicago where he spent his early working years was a hotbed of Progressive reform, led by Addams, the economist Thorstein Veblen, the philosopher John Dewey, and the writers Theodore Dreiser, Lincoln Steffens, and Upton Sinclair. Wright was intimately familiar with the Progressive education movement centered around Dewey and the historian Charles Beard, who together cofounded the New School for Social Research in New York, where Wright spoke. Dewey visited Wright at the Arizona compound built by the fellowship, Taliesin West; Beard, writing in 1934, could have been enunciating Wright's contemporary architectural vision (if not his political one) in these words: "The next America would be a collectivist democracy—a worker's republic—one vast park of fields, forests, mountains, lakes, rivers, roads, decentralized communities, farms, ranches, and irrigated deserts . . . a beautiful country—homes beautiful; communities and farms beautiful; stores and workshops beautiful. . . . Sheer Utopianism, my masters will say . . . but let it be clearly understood then that there are elements of Utopianism in all of us."

The Chicago Progressives were active participants in the growing decentralist movement in the country, part of a broad trend in urbanist thinking in the 1920s and '30s that saw urban density as a core factor in the breakdowns of those years. Addams chaired a public housing association in Illinois, and Veblen, Dewey, and Beard played a role in national debates on the issues. The influential writer Lewis Mumford forecast a "fourth migration" in the history of the city—toward a decentralized urban-rural mix not unlike Wright's vision. Mumford joined the architects and organizers Clarence Stein, Henry Wright, and Benton MacKaye in founding, in 1923, the Regional Planning Association of America (RPAA), which would go on to plan and build the model town of Radburn, New Jersey, in 1929, a garden city–inspired town organized around cars, and help to guide President Roosevelt's housing programs during the New Deal. Wright was aware of these efforts and in occasional communication with the activists, but collaborations with them proved elusive.

. . . .

A more present model for Frank Lloyd Wright was Henry Ford, the patriarch-saint of the mass-produced automobile, widely viewed in America as a hero for harnessing his innovations in machine production to a populist, decentralist social program. Before Ford, the auto was a plaything of the rich. His tough, affordable Model T had liberated the American farmer from the labor and tedium of work and transport by horses, mules, and carts—one of the few positive developments in an agricultural landscape buffeted by tumbling commodity prices and competition—and helped to slow the drift of failing farmers to the cities. Ford had built his first car in 1893, the same year Wright left Sullivan to open his own office, and both men's political and social views, maintained throughout their lives, were rooted in the populist-agrarian groundswells of that decade. In 1918, Ford the industrialist declared: "I am a farmer. . . . I want to see every acre of the earth's surface covered with little farms, with happy, contented people living on them." His means to achieving this goal echoed Wright's words: "Plainly . . . the ultimate solution will be the abolition of the City, its abandonment as a blunder. . . . We shall solve the City problem by leaving the City." Ford had been actively working on his own plans for a new kind of decentralized anticity, building rural factories and launching an effort to construct standardized worker housing near his Dearborn, Michigan, tractor plant, where 250 model Ford Homes were put up in 1919 and 1920. On December 3, 1921, Ford brought Thomas Edison and their wives by railcar to Florence, Alabama, a small community near Muscle Shoals on the Tennessee River where the federal government was planning to build a hydroelectric dam to power a nitrite plant to produce fertilizer, to free the United States from foreign dominance of that industry. The two inventors were credited with redeeming rural America through cars, roads, electrification, and Edison's movies, bringing the advantages of urbanity to small towns and the countryside. At Florence, they discussed Ford's plan for "75 Mile City," a mix of small factories, workshops, residential areas, and farms stretching for 75 miles along both sides of the river. Ford's vision rested on combining industry with agriculture: nearby mines would be linked to smelters and factories via barges and highways; hydroelectricity would light the factories and homes; and in summertime, workers would be given time off to grow some crops, making 75 Mile City fully integrated and self-sufficient. Ford's proposal, made public in 1922, was trumpeted by the national press, giving land speculators the cue to inflate local property prices, dooming the government's interest in the project.

Ford didn't give up on his ideas, keeping them alive in speeches and a 1926 book, *Today and Tomorrow*. Wright echoed this title in his own later article introducing Broadacre City in *American Architect*, calling it "Today . . . Tomorrow." In his 1930 Princeton lectures, he credited Ford as a predecessor:

> Even that concentration for utilitarian purposes we have just admitted may be first to go, as the result of impending decentralization of industry. It will soon become unnecessary to concentrate in masses for any purpose whatsoever. The individual unit, in more sympathetic grouping on the ground, will grow stronger in the hard earned freedom gained at first by that element of the city not prostitute to the Machine. Henry Ford stated this idea in his plan for the development of Muscle Shoals. . . .
>
> Even the small town is too large. It will gradually merge into the general non-urban development.

. . . .

Ford's vision would seem to have been shared by the Roosevelt administration, seeking to cope with masses of unemployed huddling in the cities: it planned to resettle them in the countryside from which they'd fled by means of intelligently conceived housing schemes merged with rural industry and farming. In 1933, during Roosevelt's "first hundred days" the Division of Subsistence Homesteads was formed in the Department of the Interior, which created 34 small communities of part-time farmers around the country, before being absorbed into the new Resettlement Administration in 1935. Wright's Subsistence Homestead in Broadacre City more than likely was inspired by these efforts. This second agency hired members of the RPAA and used its ideas in planning the three "greenbelt" towns it completed: Greenbelt, Maryland (near Washington, DC), Greenhills, Ohio (near Cincinnati), and Greendale, Wisconsin (near Milwaukee). Also in 1933, the president signed the Tennessee Valley Authority Act, creating the massive TVA agency to build dams on the rivers of the Tennessee Valley to bring electricity and industry to an impoverished region suffering doubly under Depression conditions. One of its three directors, Arthur Morgan, could have been Ford or Wright speaking when he envisioned "a valley inhabited by happy people, with small hand-work industries, no rich centers, no rich people." In partial fulfillment of Ford's plan, the TVA's first dam, Norris Dam, was built at Muscle Shoals, and some of its workers were housed in the planned town of Norris Village, a little conventional burg of historicist houses on curving streets, a low-end version of Olmsted's romantic suburbs, which fell far short of 75 Mile City and was a huge disappointment to Wright. Initially he was not recruited to the government's effort. When invited to meet with John Lansill, the head of the Suburban Division of the Resettlement Administration, Wright asked that the agency halt its greenbelt towns project and give its $100 million budget to Wright to build "the finest city in the world"—Broadacre City—with "no interference." Lansill explained that the government's programs worked with teams, but Frank Lloyd Wright was not a team player, unless he was in charge of the team, as he was with his fellowship. "Had Wright been willing to work within the liberal guidelines imposed on the rest of the planning teams," Lansill recalled, "the Resettlement Administration would have considered Broadacre City." Instead, the architect disclaimed "all public and private housing in America and never again communicated with the Suburban Division."

What has been the legacy of Broadacre City? The model was seen by 40,000 people during the one-month New York exhibition before traveling to the Wisconsin State Historical Society in Madison, Kaufmann's department store in Pittsburgh, and finally the Corcoran Gallery in Washington, DC, where it was publicized by two long pieces in the *Washington Post* and shown to scores of government officials, department heads, congressmen, transportation bureaucrats, and engineers. The Georgia landscape architect and city planner Charles Aguar wrote: "No planning proposal has ever had as much exposure or influence as Wright's Broadacre City, due in large measure to the articulation crafted into the models and the quantity and quality of publicity generated through their exhibition." Tens or hundreds of thousands more read about it in magazines and books. Yet the massive government efforts at rural city-making in ensuing years veered away from Wright's version of decentralization toward a more centralized and architecturally conventional program.

Wright himself was the first to say that achieving Broadacre City in the way he intended it was impractical: it couldn't be built piecemeal, and to fully realize it would require the abolition of cities as they were known. "I am not guilty of offering a plan for immediate use," he quipped. One critic among many called it "a naive concoction of adolescent idealism." Another, Norris Kelly, thought that achieving Broadacres "would require the abrogation of the Constitution of the United States, the elimination of thousands of government bodies from the make-up of the state, the confiscation of all lands by right of eminent domain but without compensation, the demolition of all cities and therewith the obliteration of every evidence of the country's history, the rehousing of the entire population, the retraining of millions of persons so as to enable them to be self-sustaining farmers, and other difficulties too enormous to mention. As a practicable program it does not even deserve discussion."

Architecturally, the Broadacre models generated important building types that reappeared in Wright's later work: the small apartment towers that he dubbed Saint Mark's Tower became the Price Company Tower, in Bartlesville, Oklahoma; the Pew House ("typical home for sloping ground" in Broadacre City) was the prototype of his Marin County Courthouse in California. His experimentation on the project pointed the way to other major works of his late career: Fallingwater, the Johnson Wax Building, Taliesin West, and the Guggenheim Museum in New York. These were expensive, exquisitely wrought custom structures, which found few serious imitators. What would prove far more influential was the lowly Broadacre Subsistence Homestead, to be assembled by its owner from prefabricated components. Wright had had experience with building smaller houses on budgets and using innovative construction techniques since the 1880s, and had launched a prefab venture for building homes in 1915: American System Redi-Cut Structures, of which several were built in Milwaukee, in 1916. Prefabrication was a long-standing dream of modernist architects, especially in Europe, and by the 1920s, American designers and builders longed to use Ford's standardized Model T manufacturing methods for housing. Henry R. Luce, the idealistic young editor of *Time* and *Fortune* magazines, eagerly promoted the idea, even buying *Architectural Record* in 1932 to disseminate it inside the profession. Wright was on point with his 1932 Little Farms Unit for Walter Davidson, followed by the Subsistence Homestead, drawn in November 1934. It was the first of a long line of completed Usonian Houses, including houses in Kansas, South Dakota, Wisconsin, and Florence, Alabama, near Muscle Shoals, for the Rosenbaum family, completed in 1940. Low and horizontal, with flat, cantilevered roofs, and overhangs Wright christened "carports," the Usonians had open-plan living and dining rooms adjoining a small kitchen, built-in furniture, and generous openings to outdoor spaces, frequently enclosed by an "L" plan. Using combinations of plywood sandwich walls and brick, Wright developed a Usonian "grammar" and Standard Detail Sheet to keep costs down, though finished costs generally exceeded the budget significantly. The Rosenbaums gave Wright a budget of $4,000, but ended up paying $10,000.

During the war the government enthusiastically took up prefabrication, certainly with Wright's ideas in mind. The TVA's Fontana Dam Village of 1940 was the first prefab community in the United States, with homes built of plywood in two parts at a Muscle Shoals factory, then trucked in to the site and erected for dam and defense plant workers, each home costing $2,000. By war's end, one quarter of all US housing built was factory made, totaling 200,000 units.

Afterward, Wright tried another prefab venture, this time with architect and builder Marshall Erdman in Madison. Only two houses were built—according to Erdman, due to Wright's insistence on details that couldn't be built economically: "He simply did not know how to do prefabrication." After repeated failures by the 1950s, Frank Lloyd Wright acknowledged that his houses were suitable only for the "upper middle third of the democratic strata of our country."

His Usonians nevertheless were widely publicized and copied, providing a high-architecture template for what would become the most common detached home type of the postwar suburban building boom, the ranch house. Another legacy of them, and one Wright would have been horrified by, was the "mobile home" and its proliferation across the American landscape. He may not have been its inventor, but was without doubt an important influence, both in the design and in the values that made it attractive to millions of people: low cost, expandable, suited to cheap, semirural sites—thus, as the sum of these, promising the kind of independence that Wright believed was his architecture's highest value. It was validation of his theory that the automobile would utterly transform the city, by "opening" the "door of the cage" and giving people "the means to go." Go they did, to exploding suburbs and even beyond, into the fringes of urban areas where the loose organization and low densities of suburbs were lacking. Another of Wright's Broadacre inventions, the Roadside Market, should be counted as a precursor of the ubiquitous drive-in, a key building type in enabling a new kind of radically decentralized anticity.

If the ranch house was the practical architectural expression of Wright's pioneering ideas from Broadacre City, his theories of organization for the anticity were perfected by a growing cadre of other architects and builders working to supply housing for the postwar boom. Nowhere did this happen faster or more creatively than in Southern California, before, during, and after World War II, where extraordinary population growth and high rates of car ownership met massive defense spending in a region already shaped by a decentralized settlement pattern. In this period, California won more defense dollars than any other state, drawing huge numbers of workers, employment increased by 75 percent from 1939 to 1943. In Southern California, by 1941 nearly half the region's manufacturing jobs were in aircraft manufacturing, and 13,000 new industrial workers arrived in Los Angeles every month. By 1943, there were 400,000 defense workers in Los Angeles County—one quarter of the entire workforce, and the multiplier effect was huge: LA County payrolls grew from 900,000 in 1940 to 1,450,000 in 1943—a 60 percent gain, with the number of manufacturing workers tripled. They earned some of the highest wages in the nation: 141.2 percent of the per capita national average income. Existing housing couldn't begin to accommodate the need, nor could existing development methods.

Before the war, development of worker housing proceeded largely on an ad hoc basis of subdividing vacant or agricultural land into small, inexpensive parcels, where owners would either build their own structures or hire contractors to do so for them—often using partly prefabricated house "kits," like those offered by Pacific Ready Cut Homes, whose bungalow cost $2,750 in 1924. The resulting landscape was a mostly unplanned, accretive, blue-collar suburbia, on unincorporated land near outlying industrial centers, oil fields, or airfields, with few or no public services. But as the number of war workers grew—by 1943, the Lockheed and Vega cluster of plants in the Burbank area, including ancillary suppliers, employed 72,000 people, with another 15,500 elsewhere in the Valley—the old model of ad hoc building was overwhelmed. Defense manufacturers partnered with federal authorities and private developers to apply the same mass production methods their industry used to build entire communities from scratch. Critical to their efforts were a raft of new federal financing tools: the 1941 Lanham Act, which provided $1.3 billion for war-worker housing in areas of shortages, and Title VI of the Housing Act, guaranteeing federal loans up to 90 percent of project value for private housing developments in shortage areas. In 1940, Douglas Aircraft broke ground on a tract of 1,100 houses on unincorporated Los Angeles County land near Long Beach, about two miles from its plant then gearing up to build B-17 bombers. In Santa Monica, near another Douglas plant, Westside Village went up, a tract of 885-square-foot homes carefully optimized for space, including a kitchen with modern appliances and built-in storage to eliminate the pantry facing the backyard, in up-to-date suburban fashion. The builders used a continuous-flow assembly line, putting together parts provided by nearby suppliers, enabling a high volume at great speed, high quality, and a low price: $3,290, for $190 down and $29.90 a month. Not far from the North American and Douglas plants at Mines Field, which would later become Los Angeles International Airport, a tract of 3,230 homes, for 10,000 people, took shape from 1942 to 1944. These developments constituted an industrial revolution in home building, a decade before Levittown.

The postwar Los Angeles that emerged was a decentralized, regional city, with its nodes sown from the principal aircraft plants and grown into surrounding, purpose-built communities: North Hollywood/Burbank/Glendale (Lockheed, Vega), Santa Monica/Mar Vista/Culver City (Douglas, Hughes Aircraft), Inglewood/Westchester/El Segundo (North American Aviation, Douglas El Segundo, Northrop), Downey (Vultee), and Long Beach/Huntington

Beach (Douglas Long Beach). These spreading clusters traced a ring roughly 15 miles from Los Angeles City Hall, linked by an emerging system of freeways, primed again by federal funding, and serviced by regional shopping centers surrounded by enormous surface parking lots built by developers. Modern LA, and with it the growth model for countless other communities and cities across the globe over the next decades, had taken shape.

The next problem to be tackled after planning and design was, as in Broadacre City, governance. Again, the critical advance came in Southern California, in the aircraft suburb of Lakewood. In 1950, next door to the neighborhood built for workers at the Douglas Long Beach plant, another housing development was planned, the biggest ever undertaken in America to that point. Platted on 3,500 acres of farmland, Lakewood would comprise 17,500 houses, each 1,100 square feet, in one of several floor plans carefully sequenced so that no two identical homes were adjacent. With $100 million in federal mortgage financing, the developer set up a full-scale industrial assembly line to build houses: 100 a day at full tilt, 500 a week, with the total finished in three years. When the sales office opened, 25,000 people were waiting.

Part of the appeal of Lakewood and places like it in the 1950s was its separateness: not only from the central city that people left to follow jobs moving to the periphery, but also from other people, especially certain kinds of other people. The LA area had, since early in the twentieth century, advertised itself as "the white spot" of America to attract white migrants and had pioneered the use of the private homeowners' association—in effect, private governments responsible for common property—to enforce separation through deed restrictions, exclusionary zoning, and design regulation. The first was the Los Feliz Improvement Association, founded in 1916 in that Los Angeles neighborhood, and was quickly copied. "Their overriding purpose," according to author Mike Davis, "was to ensure social and racial homogeneity." By 1941, most of the city itself, plus neighboring cities on the Westside, in San Gabriel Valley, and in Pasadena were closed to nonwhites. But this supposed homogeneity had been eroded by an influx of Southern blacks during the Depression and then the war. Drawn by the promise of defense jobs, blacks poured in: as many as 10,000 per month during the height of the buildup. They remained confined to certain areas, mostly in the central city and outlying industrial districts—approximately 5 percent of the city's residential area. By war's end their population had doubled within these tight bounds and had spilled over into downtown's Little Tokyo, its Japanese residents having been shipped to internment camps. Residents of adjoining neighborhoods responded by tightening racial restrictions, or moving out. The 1948 Supreme Court decision outlawing housing discrimination began the slow dismantling of color barriers and accelerated the process of white flight from central city neighborhoods to new, peripheral suburbs, which even after 1948 were advertised implicitly as restricted to white people. The sales brochures for Lakewood proudly trumpeted its "race restrictions," keeping it a "100% American Family Community," "the white spot" of Long Beach.

In 1953, the city of Long Beach announced its intention to annex Lakewood. Some Lakewood residents responded with a drive to incorporate the development as a municipality, leading to a very public battle, ending with Lakewood's incorporation in 1954. The decision to incorporate was swung by an innovative agreement reached with the Los Angeles County government to contract the same services the county provided unincorporated areas: road building and maintenance, the health department, building inspection, libraries, school services, the animal pound, tax collection, the fire department, and the sheriff for law enforcement. This arrangement, known as the Lakewood Plan, allowed small communities that would have had no chance of affording cityhood's services and payrolls to put up walls and exercise control over zoning while maintaining low tax rates. As further inducement, in 1956, the state of California granted cities 1 percent of local sales tax. Developments like Lakewood that had been built around a shopping center suddenly had a coveted tax base, and a replacement for the diversified downtown business district of traditional urbanism. The Lakewood Plan spread like wildfire: two years later, 4 more LA County communities became cities on the plan; in 1957, 5 more followed and 23 were involved in the process. Thirty cities incorporated in the space of two years, when in the previous 106 years, a total of 45 had become cities in the county. All around Lakewood, miniature municipalities sprouted: Pico Rivera, Paramount, Montebello, Cerritos, Bellflower, Bell, Bell Gardens, Hawaiian Gardens, Maywood, Cudahy, Commerce, and so on; the San Gabriel Valley followed suit, filling south LA County with grids of streets and houses covering hundreds of square miles.

Since 1954, all but one of the cities incorporated in LA County have adopted the Lakewood Plan. Eighty percent of cities incorporating in California adopted it, bringing the total number of California cities based on it to 30 percent. It is emulated all over the United States. The arrangement confers real advantages by providing services at reduced tax rates while keeping control of the shape, and therefore makeup, of the city at home—home rule at its simplest. Not all such places were conceived of as, or are, racially exclusive, but many have been, and remain so. When color barriers broke down after 1948, many of the first generation of restrictive homeowner associations (HOAs) failed in their intent, but were succeeded by a new generation of Lakewood-style cities forming a second-ring suburban "white wall" around evacuated central cities like Detroit, Saint Louis, and New Orleans, maintaining racial separation through residential and economic segregation. By the 1990s, there were 16,000 HOAs in California alone—an increasing percentage of them gated communities, which simply render their fencing functions visible with actual fences and private-security measures.

From a plane, the majority of the land area in and near US urban regions can clearly be seen to be either suburban sprawl of the "cul-de-sac" developer homestead type, or pseudo-rural and semirural trailer-and-McMansion sprawl. Can this pattern be traced to Frank Lloyd Wright? Certainly, the exclusionary intents and effects of HOAs are a grotesque exaggeration of his bedrock principle of the independence of the patriarchal family and the individual. There is nothing in his writings to support the charge of racism. And yet it is the same principle: of separation through strict spatial control. And sprawl is exactly what Wright was so carefully planning. Compare it with this definition of sprawl by the urban planner and architect John Dutton: "diffuse, de-centered, without clear boundaries, and car-dominated." The view from the window of the plane would seem to demonstrate that Americans, while apparently ignoring it, in fact have nearly achieved Wright's vision of an unbounded landscape of automobile-enabled, pseudo-rural settlement blanketing the continent from coast to coast, limited only by mountains, bodies of water, and deserts. There is irony in this: while Wright saw his Subsistence Homestead as a means to partial economic independence, inhabitants of actual suburbia depend for their jobs and mortgages on huge, outside corporations and banks, and on big government for their public services. And the suburbs have had to be massively subsidized, since decentralized sprawl is a net drain compared with traditional cities on infrastructure (roads, water, sewerage, and power transmission) and resources (for driving; for energy to heat, cool, and power enormous houses and maintain enormous landscapes; and for the greenhouse gases that go with them). The heavily suburbanized United States has 4.5 percent of the global population but consumes 25 percent of global resources. This fact provoked the utopian architect Paolo Soleri—who for 18 months in 1947–48 was a Taliesin apprentice, before being expelled by Wright—to take the master to task in a 1992 interview: "That's the big, big failure of Frank Lloyd Wright. . . . That's a tragedy, more than a failure . . . because it encouraged, it glamorized the notion of the single home . . . the family home. And the spread out, the two-dimensional spread out, and consumerism. There's nothing as consuming as suburbia. It's a . . . colossal engine for consumption. So—I'm sure that if Mr. Wright was alive now he would have changed his rationale about Broadacre City."

4. Homesteads

Field Guide: Sprawl

Diagnostic Characteristics/Diagnostics:

- Homesteads are single-family houses on land outside city centers, with pseudo-rural qualities, suggesting agrarian self-sufficiency, though they are in reality not farms, but typically bedroom communities for urban areas, or homes for retirees. Usually in areas set aside for low-density residences. Either unplanned, as in unincorporated areas outside municipalities, or in planned communities or entire cities conceived as homestead zones.
- These areas are radically decentralized, have no functional center, virtually no mass transportation, and are completely car-dependent: sprawl.
- The architectural style in these areas is nonspecific, from historically themed subdivisions and carefully controlled gated communities, to random, commercial vernacular.
- Building form is predominantly detached residential, but can include pockets or mixes of multiple types, heights, and styles, accommodating many uses, from strictly residential to industrial to agricultural to golf courses to prisons, which are preferentially sited in peripheral or quasi-rural sprawl zones.

Examples:

- Subdivisions, industrial and shipping complexes, on city edges or on highways.
- Peripheries of most American cities, from the first- and second-ring suburbs inside city limits to suburban municipalities and unincorporated areas outside city limits: Detroit, Michigan; Indianapolis, Indiana; Bozeman, Montana.
- Dominant form in most cities largely built in the twentieth century: Phoenix, Arizona; Orlando, Florida; Santa Clarita, California.

Images:

Broadacre City renderings, Frank Lloyd Wright.

Rosenbaum House (1940), Florence, Alabama. Architect: Frank Lloyd Wright.

Housing development (1972), Florida.

Subdivision (1973), Maryland.

Subdivisions (1975), off Mulholland Drive, Santa Monica Mountains, Los Angeles, California.

Subdivisions (2009), Rio Rancho, New Mexico.

Chapter [5]

Corals

Jane Jacobs, Andres Duany, and the Self-Organizing City

There is a quality even meaner than outright ugliness or disorder, and this meaner quality is the dishonest mask of pretended order, achieved by ignoring or suppressing the real order that is struggling to exist and to be served.

—Jane Jacobs, *The Death and Life of Great American Cities*

If we wish to tie up with our colonial tradition we must recover more than the architectural forms: we must recover the interests, the standards, the institutions that gave to the villages and buildings of early times their appropriate shapes. To do much less than this is merely to bring back a fad which might as well be Egyptian as "colonial" for all the sincerity that it exhibits.

—Lewis Mumford, *Sticks and Stones*

It is an observable phenomenon that the more comfortable, self-sufficient, and grand the private home becomes, the less time its residents are likely to spend in the public forum. As trends have been established in the United States, most notably private affluence and public poverty, people tend to overindulge in their homes to offset the deficiency of the broader context of the community in which they live.

—Richard Sexton, *Parallel Utopias*

During the 1950s, New York City's "master builder" Robert Moses was at the height of his powers, near to completing his grand project of remaking the metropolitan region in the image of Le Corbusier's Radiant City: with towers-in-parks standing where "blighted" districts had been, and the tristate area of New York, New Jersey, and Connecticut linked together by an integrated system of expressways, bridges, and tunnels. He had moved hundreds of thousands of people from their homes, cleared thousands of acres, and torn up thriving neighborhoods, all in the service of progress and modernization. In the words of historian Hilary Ballon: "Moses had more power over the physical development of New York than any man had ever had or is ever likely to have again." Moses understood this power, directly comparing himself to his predecessor Baron Haussmann, whose "dictatorial talents" he wrote, "enabled him to accomplish a vast amount of work in an incredibly short time, but they also made him many enemies, for he was in the habit of riding rough-shod over all opposition." His favorite technique for disarming opponents was to vilify them as out of touch with the needs of the bulk of the people, as he had in the 1920s in order to ram the Southern State Parkway through a wealthy residential area on Long Island, lambasting the homeowners as "a few rich golfers." Claiming the populist mantle of "the People's Administrator," Moses, ruling like an autocrat, was effectively unstoppable.

Moses's plan for completing the expressway connections for the region included the Lower Manhattan Expressway, which would connect the Holland Tunnel on Manhattan's lower West Side to the Manhattan and Williamsburg bridges that crossed the East River to Brooklyn, thereby providing a car-and-freight corridor between

New Jersey and Long Island through a badly congested, still partly industrial portion of Manhattan. The elevated road would cut across the island along Broome Street with a right-of-way up to 350 feet wide, requiring the bulldozing of a broad swath through SoHo, Little Italy, the Bowery, and Chinatown, in the process displacing 2,000 families and 800 businesses, and razing eight churches, a police station, and a park. LOMEX, as the expressway was called, would link up with Midtown to the north via Fifth Avenue, by extending that thoroughfare southward through Washington Square Park as a four-lane roadway that would go on to connect to the expressway at Broome Street. The park, which occupies a roughly 10-acre rectangle at the center of Greenwich Village, had since the nineteenth century been the neighborhood's heart, celebrated for its 1892 triumphal arch by Stanford White, wide fountain, shade trees, and benches, providing an oasis for residents, children, tourists, and the faculty and students of New York University, which partly surrounds it. A narrow carriageway runs through it from north to south, passing under the arch, but is normally closed to traffic. As early as 1935, Moses had identified the park and its adjacent, small streets as a bottleneck slowing automobile traffic through the city. His plan to widen the peripheral street had been blocked by complaints from the university. But by the 1950s, with the enormous momentum of a decades-long career behind him, his ambition had grown: not only did his plans include a roadway through the center of the park, to be Fifth Avenue South, but he targeted 12 acres of mixed industrial buildings and apartment blocks stretching south of Fourth Street as "blighted," proposing to clear them and build a Title I project in their place, Washington Square Southeast.

When word of his plans leaked out in the spring of 1952, local residents protested. One wrote to the commissioner: "The democratic way is to allow the people of the community to have a voice in its projected use. We urge you to schedule public hearings in which we may participate before you proceed. We cherish the right to participate in the planning of our community." But Moses was having none of this vision of participatory democracy, and shot back: "It must be obvious that this cannot be settled by a mass meeting." Finding themselves stonewalled, a group of residents, many of them mothers whose children had few other options for playing outdoors, attended a meeting of the city's Board of Estimate, which they knew to have ultimate permitting authority over Moses's plans, and won a temporary suspension of the project. The *New York Times* registered the disbelief that Moses surely felt: "Project That Would Put New Roads in Washington Square Park Upset by Women."

One of those women was Shirley Hayes, a once-aspiring actress and mother of four, who threw herself into opposition, organizing the Washington Square Park Committee, gathering signatures, and insisting on the special value of the park: as a safe place for kids to play, for adults to socialize, and for the very diverse inhabitants of the neighborhood to mingle with one another. The Village wasn't "blight," but a functioning city, and home. Greenwich Village was unique in Manhattan: with narrow streets and short blocks, small neighborhood parks, mostly low- to midrise buildings occupied by a mix of industry, retail businesses, offices, and residences, and a population unusually mixed for the island, with sizable communities of Italians and Irish. The West Village, laid out in the eighteenth century before the grid plan of 1811, retained its eccentric street layout, with crosstown streets kinking northward to the west of Sixth Avenue, and the crossing blocks kinking south, posing unique navigational challenges—for example, where West Fourth Street intersects with West Thirteenth Street.

Where some other civic groups talked of compromise, perhaps by downsizing the roadway, Hayes and the committee remained adamantly opposed to any road at all through the park. They had seen the results of Moses's work on other parts of the region and believed that any concessions to the designs of Urban Renewal would soon become an opened floodgate, and they resolved to stand firm. They were politically savvy, understanding that the Board of Estimate was made up of elected officials—Mayor Robert Wagner, the five borough presidents, and two members elected at large for the entire city—and as such were politicians responsive to organized voters. They packed meetings, organized protests in parks, and used the media—especially the "new" media, above all the alternative weekly *Village Voice* newspaper, which began publication in 1955—and found in the controversy an ideal rallying point for a discussion about the future of New York. The *Voice* editor Dan Wolf editorialized: "It is our view that any serious tampering with Washington Square Park will mark the true beginning of the end of Greenwich Village as a community. . . . Greenwich Village will become another characterless place."

Such protestations failed to persuade Moses or his allies in the big business world who saw the roadways as critical to modernization. Moses's aid Stuart Constable dismissed the residents out of hand, with a curious slur: "I don't care how those people [in Greenwich Village] feel. . . . They're a nuisance. They're an awful bunch of artists down there." But, instead of intimidating the dissidents, the fight against Robert Moses became a cause célèbre, galvanizing local residents and attracting prominent outsiders to weigh in, such as the anthropologist Margaret Mead, and Lewis Mumford, who called the city's plan "civic vandalism." William H. Whyte, a writer at *Fortune* magazine whose 1956 book *The Organization Man* (about the exact corporate executives allied with Moses) sold 2 million copies, said that Urban Renewal was "planned by people who don't like cities," and that the resulting projects were guilty of "vast redundancy . . . no sense of intimacy or of things being on a human scale." At a June 1958 gathering in support, Columbia University planning professor Charles Abrams praised the opposition as an epochal "revolt of urban people against the destruction of their values; of the pedestrian against the automobile; the community against the project; the home against the soulless multiple dwelling; the neighborhood against the wrecking crew; of human diversity against substandard standardization." He damned what he called the dominance of traffic engineers and the "heedless destruction that has been the theme of the slide rule era 1935–1958"—and lauded the "rediscovery of what is good in our cities"—that is, smallness, diversity, and community.

In response to calls for a smaller roadway, Moses didn't blink, but doubled down, increasing the planned width to 48 feet. Another of the Greenwich Village mothers who helped in the years-long struggle by collecting signatures, demonstrating, and speaking was Jane Jacobs, who lived with her husband and three children in the West Village. She recalled only one face-to-face encounter with Robert Moses, at a raucous Board of Estimate meeting: "He stood up there gripping the railing, and he was furious at the effrontery of this [opposition]." Jacobs remembered him shouting: "'There is nobody against this—NOBODY, NOBODY, NOBODY, but a bunch of, a bunch of MOTHERS!' And then he stomped out."

In 1958, an umbrella group was formed, the Joint Emergency Committee to Close Washington Square Park to Traffic, with Jacobs as its chair. They collected 30,000 signatures for an "experimental" plan to close the park temporarily to all traffic—a gambit that Moses was prepared to accept if only to prove that it would make traffic congestion in the neighborhood unbearable. The committee took advantage of political leverage, appealing publicly to the entrenched Democratic Party boss, Carmine DeSapio, who lived on Washington Square, and was facing a serious challenge from Ed Koch (the future mayor). DeSapio sided with the road opponents. On September 18, the Board of Estimate voted for the closure. On November 1, 1958, a ceremonial ribbon was stretched across the road—not cut—symbolically protecting the park. DeSapio held one end, while Jane Jacobs's daughter Mary, "representing the children," held the other. Professor Abrams, mock-seriously, would comment: "It is no surprise that, at long last, rebellion is brewing in America, that the American city is the battleground for the preservation of diversity, and that Greenwich Village should be its Bunker Hill. In the battle of Washington Square, even Moses is yielding, and when Moses yields God must be near at hand."

The victory held, as drivers continued to find ways around the park, as they always had. It marked a turning point, showing other communities in the path of Urban Renewal that well-organized resistance could succeed, such as those neighborhoods in the path of Moses's LOMEX project, which, with the help of Jacobs and a new group, the Joint Committee to Stop the Lower Manhattan Expressway, managed to block it. By 1960, the project's cost had reached $100 million, but with 90 percent due to come from the federal government, Robert Moses didn't finally let go of the plan until 1969. Nationwide, Urban Renewal only ramped up through the 1960s and beyond, swinging its "meat axe," in Moses's indelible words, through city after city. After the park struggle, Shirley Hayes faded from public view. She wasn't even mentioned in Robert Caro's 1974, 1,250-page, Pulitzer Prize–winning biography of Moses, *The Power Broker*, which etched him in the nation's conscience as a dictatorial and destructive figure. But Jane Jacobs, whom Robert Fishman described as "a foot soldier more than a leader in the actual battle," and whose diminutive, owl-like appearance must have made her seem less formidable than she was —went on to alter the course of history.

Jane Jacobs was born Jane Butzner in 1916, in Scranton, Pennsylvania. After high school she worked as an unpaid assistant to the local newspaper's women's page editor, before moving to New York during the Depression with her sister, finding employment as a stenographer and a freelance writer, and settling in Greenwich Village. She took classes uptown at Columbia's School of General Studies, exploring a kaleidoscopic range of interests, including zoology, geology, political science, law, and economics. During the war she worked for a business magazine and then as a writer for the US Office of War Information. In 1944, she married Robert Hyde Jacobs, Jr., an architect, and in 1947 the couple bought a modest three-story building in the West Village, at 555 Hudson Street between West Eleventh and Perry streets, for $7,000. In 1952, Jacobs took a job at Henry Luce's *Architectural Forum*, covering, among other issues, Urban Renewal. She wrote a critical piece on Philadelphia's equivalent to Robert Moses, Edmund Bacon, which brought her to the attention of William Whyte, who got her an assignment for *Fortune* magazine for a 1958 piece entitled "Downtown Is for People." It was reported to have enraged the magazine's publisher, General Charles "C. D." Jackson, who demanded of Whyte, "Who is this crazy dame?" Jacobs was routinely dismissed in many quarters as a "housewife" lacking expertise, yet her systematic and trenchant critiques of modernist planning landed her an invitation to speak at a 1959 planning conference at Harvard University, which in turn led to an award from the Rockefeller Foundation to put her thoughts into book form. The result, *The Death and Life of Great American Cities*, published in 1961, was a sensation.

In its first line she announced her uncompromising, unambiguous intention: "This book is an attack on current city planning and rebuilding." So began a fairly cheerful, drily witty, levelheaded, but relentless excoriation of the planners and their mind-set. Jacobs stated flatly that the profession was a "pseudoscience," in which "years of learning and a plethora of subtle and complicated dogma have arisen on a foundation of nonsense," starting with its economic justifications. "The economic rationale of current city rebuilding is a hoax," she went on, requiring billions in taxpayer money, plus the "vast, involuntary subsidies wrung out of helpless site victims," who were removed or otherwise affected by Urban Renewal schemes. All supposedly to achieve illusory "increased tax returns"—"a mirage, a pitiful gesture against the ever increasing sums of public money needed to combat the disintegration and instability that flow from the cruelly shaken-up city." The miserable results of this false accounting spoke for themselves, she wrote:

> There is a wistful myth that if we only had enough money to spend—the figure is usually put at a hundred billion dollars—we could wipe out all our slums in ten years, reverse decay in the great, dull, gray belts that were yesterday's and day-before-yesterday's suburbs, anchor the wandering middle class and its wandering tax money, and perhaps even solve the traffic problem.
>
> But look at what we've built with the first several billions: Low-income projects that become worse centers of delinquency, vandalism and general social hopelessness than the

> slums they were supposed to replace. Middle-income housing projects which are truly marvels of dullness and regimentation, sealed against any buoyancy or vitality of city life. Luxury housing projects that mitigate their inanity, or try to, with a vapid vulgarity. Cultural centers that are unable to support a good bookstore. Civic centers that are avoided by anyone but bums, who have fewer choices of loitering places than others. Commercial centers that are lackluster imitations of standardized suburban chain-store shopping. Promenades that go from no place to nowhere and have no promenaders. Expressways that eviscerate great cities. This is not the rebuilding of cities. This is the sacking of cities.

The pseudoscience's second foundation was the notion of blight. The language of pathology was off the mark, she said: "Medical analogies, applied to social organisms, are apt to be far-fetched, and there is no point in mistaking mammalian chemistry for what occurs in a city." Nevertheless, Jacobs reversed the medical analogy to make her point, comparing orthodox modernist planning to the "elaborately learned superstition" of nineteenth-century medicine, "when physicians put their faith in bloodletting, to draw out the evil humors that were believed to cause disease." The scientifically sound medical analogy to cities was instead, for Jacobs, that "sick people needed fortifying, not draining." The dogma of "blight" was based on a series of a priori beliefs: that commerce and industry mixed with housing led to pollution, congestion, and moral pathologies; that high housing densities equaled dangerous overcrowding; that the absence of substantial amounts of green space forced children to "play in the streets," or "the gutter," which was by definition unhealthy and unsafe; that narrow streets and small blocks were inefficient for vehicular circulation, "in planning parlance . . . 'badly cut up with wasteful streets'"; that old buildings presaged slum conditions of dilapidation and neglect. By these assumptions, neighborhoods like Greenwich Village or the North End of Boston must be slums: "in orthodox planning terms" they existed in "the last stages of depravity."

Though 20 years earlier Boston's North End, a small, low-rent neighborhood lying between the industrial waterfront and the older downtown, had been crowded with recent immigrants and visibly dilapidated, what Jacobs saw on a visit there in 1959 was starkly different. Many of the residents owned homes and had upgraded their buildings—from their savings, as bank mortgages were nearly impossible to get owing to the neighborhood's classification as a slum by government agencies. There was a rich variety of small businesses, including "splendid food stores," mixed together with small industrial and craft workshops. "The streets were alive with children playing, people shopping, people strolling, people talking." It had a "general street atmosphere of buoyancy, friendliness, and good health." She thought, "It doesn't look like a slum to me." Social statistics bore her impression out, in low rates of disease, child mortality, and crime. Even so, a banker friend of hers averred that it was by definition Boston's "worst slum," given its intolerable 275 dwelling units to the acre. "You should have more slums like this," she told him.

The North End along with the similar West End of Boston were being targeted for Urban Renewal, which would destroy them. Jacobs allowed that "the intentions" of the planners and architects drawing up those plans were "on the whole exemplary," seeking to improve the situation of cities in the era of suburbanization. The problem was that they paid too-careful attention to "what the saints and sages of modern orthodox planning have said about how cities ought to work and what ought to be good for people and businesses in them."

One by one, Jacobs called out the giants of twentieth-century city planning, beginning with Ebenezer Howard and his garden city. He was so convinced of the innate evil of the city, which in his frame of reference was the nineteenth-century industrial city familiar from the novels of Dickens, "his prescription for saving the people was to do the city in," by replacing it with its opposite, modeled on the preindustrial English manor and village, standing apart both in time and economic space from the city. "His aim was the creation of self-sufficient small towns, really very nice towns if you were docile and had no plans of your own and did not mind spending your life among others with no plans of their own." She charged Howard with introducing two foundational ideas of modern planning: one, of "sorting out" the city's functions, to "arrange each of these in relative self-containment," and two, of focusing on "the provision of wholesome housing as the central problem," which he defined only in terms of "suburban physical qualities and small town social qualities," not urban ones. "Virtually all modern city planning has been adapted from, and embroidered on, this silly substance."

She moved on to Howard's faithful disciples, the American Decentrists, including Lewis Mumford, Catherine Bauer, Clarence Stein, Henry Wright, and the RPAA, who, she alleged, were antistreet and antidensity, and therefore anti–other people: "The presence of many other people is, at best, a necessary evil, and good city planning must aim for at least an illusion of isolation and suburban privacy." At base, the Decentrists were, like Howard, anti-urban: "The great city was Megalopolis, Tyrannopolis, Nekropolis, a monstrosity, a tyranny, a living death. It must go." It was no wonder then that the planners failed to see the actual city, as opposed to its nineteenth-century nadir: "How could anything so bad be worth the attempt to understand it?" Real cities, she wrote, "are an immense laboratory of trial and error, failure and success." But the planners "ignored the study of failure and success in real life" and "have been incurious . . . and are guided instead by principles derived from the behavior and appearance of towns, suburbs, tuberculosis sanatoria, fairs, and imaginary dream cities—from anything but cities themselves."

One of those dream cities was Le Corbusier's towers-in-the-park, so destructively grafted into the flesh of New York. His vision came directly from the pseudo-rural garden city, she thought, and thus its obsession with "the super-block . . . the unchangeable plan, and grass, grass, grass." Its pleasant nature imagery explained people's "creeping acceptance" of the Radiant City's hyperconcentration. "If the great object of city planning was that Christopher Robin might go hoppety-hoppety on the grass, what was wrong with Le Corbusier?" The answer was his reductive but seductive misreading of the complexity of real cities: his "dream city" was "like a wonderful mechanical toy" with "a dazzling clarity, simplicity, and harmony. It was so orderly, so visible, so easy to understand . . . all but irresistible to planners, housers, designers, and to developers, lenders, and mayors too." "But as to how the city works, it tells, like the Garden City, nothing but lies." Lastly, she lambasted the City Beautiful of Daniel Burnham as "retrogressive imitation Renaissance style." She likened the Chicago fair's grand neoclassical buildings to Le Corbusier's slabs: "One heavy, grandiose monument after another . . . arrayed . . . like frosted pastries on a tray, in a sort of squat, decorated forecast of Le Corbusier's later repetitive ranks of towers in a park." The common theme of all these planning recipes was their intention of "decontaminating their relationship with the work-aday city" through repression of its diversity and the separation of its major functions. Layered over one another, these functions "merged, into a sort of Radiant Garden City Beautiful." The prime example of which was, wrote Jacobs, New York's immense Lincoln Square Urban Renewal project, which she reckoned a sterile disaster erected over the ruins of a once-thriving neighborhood wrongly condemned as blighted.

"From beginning to end . . . the entire concoction [of planning] is irrelevant to the workings of cities," Jacobs concluded. "Unstudied, unrespected, cities have served as sacrificial victims." In contrast, she defined her project as first rejecting the temptations of planning dogma's "specious comfort of wishes, familiar superstitions, oversimplifications, and symbols," and then embarking "upon the adventure of probing the real world," in order "to look closely . . . at the most ordinary scenes and events, and attempt to see what they mean and whether any threads of principle emerge among them." First, she said, one must be clear about the object: "Great cities are not like towns, only larger. They are not like suburbs, only denser." Their rules are distinct. They aren't collections of object-buildings, each with a distinct function, the provision of which will guarantee social order and good outcomes—a belief that Reinhold Niebuhr called the "doctrine of salvation by bricks alone." A city isn't just schools, housing, parks, and buildings, but the interactions between them and their inhabitants and users. Cities, insisted Jacobs, are "problems in organized complexity."

The most basic unit of study was streets and sidewalks, a city's "most vital organs." When working well, streets perform three necessary functions: safety, contact, and "assimilating" children. They do so by being used: first by people who live or work in the neighborhood, interacting with one another, and above all, keeping an eye on the goings-on, especially of children, whether their own or others'. This surveillance Jacobs called having "eyes on the street," and it is, she insisted, what makes good streets safe for the locals and for outsiders—essential because cities are, by definition, full of strangers, and strangers' safety, too, is provided more by the community—merchants, shopkeepers, parents, strollers, even bar patrons—than by police. The necessary glue is trust, slowly accreted from "networks of small-scale, everyday public life":

> The trust of a city street is formed over time from many, many little public sidewalk contacts. It grows out of people stopping by at the bar for a beer, getting advice from the grocer and giving advice to the newspaper man, comparing opinions with other customers at the bakery and nodding hello to the two boys drinking pop on the stoop, eying the girls while waiting to be called for dinner, admonishing the children, hearing about a job from the hardware man and borrowing a dollar from the druggist, admiring the new babies and sympathizing over the way a new coat faded. . . . The sum . . . is a feeling for the public identity of people, a web of public respect and trust, and a resource in time of personal and neighborhood need. The absence of this trust is a disaster for a city street. Its cultivation cannot be institutionalized.

The first goal of planners therefore ought to be "to foster lively and interesting streets," and Jacobs listed some critical factors. First, people in the main are not in cars, but on foot, and the street must be understood as a primarily pedestrian realm. Thus her description of what happens there was of dance: "the daily ballet of Hudson Street."

Second, there have to be enough people using the street, and using it "fairly continuously," which can best be achieved through a wide mix of residence, work, shopping, and nightlife—including, she took pains to defend, bars, like the famous White Horse Tavern on her block of Hudson Street, which kept "eyes on the street" into the morning hours. "To understand cities," she wrote, "we have to deal outright with combinations or mixtures of uses, not separate uses, as the essential phenomena." She warned of the deadness that comes from segregating urban life into a "series of decontaminated sortings," not least downtowns empty of people "after working hours": "This condition has been more or less formalized in planning jargon, which no longer speaks of 'downtowns' but of 'CBDs,'" central business districts, like the tip of Manhattan, which was almost completely vacated after five o'clock. "Without a strong and inclusive central heart, a city tends to become a collection of interests isolated from one another. It falters at producing something greater, socially, culturally, and economically, than the sum of its separate parts." Third, there must be a "clear demarcation between public and private." When the two are blurred what results are dangerous "project prairies" and "blind-eyed reservations," unprotectable even by fences, virtual or real. There, "the keeping of public sidewalk law and order is left almost entirely to the police and special guards. Such places are jungles. No amount of police can enforce civilization where the normal, casual enforcement of it has broken down."

Finally, Jacobs boiled her observations down into a succinct grammar: "To generate exuberant diversity in a city's streets and districts, four conditions are indispensable." One, the district must serve more than one primary function: that is, the "close-grained diversity of uses that give each another constant mutual support, both economically and socially." Two, most blocks should be short, to multiply people's choices of routes, and so opportunities for interaction, and increase usable street frontage, which expands the "pool of economic use." Long blocks, especially where residences are separated from businesses, result in "the dismally long strips of monotony and darkness—the Great Blight of Dullness, with an abrupt garish gash at long intervals. This is a typical arrangement for areas of city failure." Three, the district must mingle buildings that vary in age and condition, to allow economic diversity; as she claimed, older buildings, with lower costs to pay off, would be available for less money, allowing younger people and entrepreneurs opportunities. Four, there must be a sufficiently dense concentration of people. This, she wrote, was partly a function of condition number one—mixed primary uses drawing people at different times; but was also dependent on having sufficient housing density. She pointed out that high density and overcrowding are not the same; in fact, "overcrowding of dwellings and high densities of dwellings are always being found one without the other." The key is to have the latter without the former. Together, these are "generators of diversity" and breed success: successful streets generate successful neighborhoods, then successful districts, and then successful cities. The relationship is circular: "City diversity itself permits and stimulates more diversity." She offered no guarantees, but maintained that, if the four conditions were met, "city life will get its best chances."

What Jacobs sought was to discover the rules for self-organizing cities. While this may sound like an oxymoron, it isn't unlike seeking to understand how coral reefs—enormous, diverse, and complex structures—build up over time through the accumulation of tiny, simple organisms, each acting on its own, following clear patterns of growth and assembly, gradually adding up to the vitality of the whole. The greater the number of coral species and colonies, the greater the diversity of associated species of shrimp, fish, nudibranchs, crustaceans, mollusks, and so on—its economy—and the greater the reef's functionality and resilience. The quality of resilience is comparable to Jacobs's definition of a "successful" city neighborhood as "a place that keeps sufficiently abreast of its problems so it is not destroyed by them. An unsuccessful neighborhood is a place that is overwhelmed by its defects and problems and is progressively more helpless before them." And like the colonies making up a reef, neighborhoods shouldn't be "islanded off as a self-contained unit," as small towns isolated from others by greenbelts, as in Howard's garden city or the Decentrists' greenbelt towns, or by parks and roadways, as in Le Corbusier's Radiant City. "Successful street neighborhoods, in short, are not discrete units. They are physical, social, and economic continuities—small scale to be sure, but small scale in the sense that the lengths of fibers making up a rope are small scale."

The Death and Life of Great American Cities was received like an alarm sounding, earning a wide readership eager to understand why, even after billions spent, Urban Renewal seemed to be making the "urban crisis" worse instead of better. Inside the planning profession, Jacobs was generally dismissed as an amateur at best, and a meddling housewife at worst, including by Lewis Mumford in a 1962 *New Yorker* magazine article called "Home Remedies for Urban Cancer," in which he called her "a new kind of 'expert'" and an "able woman" who "used her eyes, and, even more admirably, her heart to assay the human result of large-scale housing," but who was unqualified to criticize the scientific expertise of professionals. The book would eventually be translated into six languages and sell a quarter-million copies. Jacobs's work was part of a larger dialogue, both national and international, about the causes of the decline of central cities, fueled by deep anxiety about the runaway growth of suburbs everywhere and particularly in the Sunbelt, which was being transformed into a collection of metastasizing, suburban city-regions like Southern California, Phoenix, Houston, and South Florida. Combined with widespread doubts about Urban Renewal, this suburban sprawl helped spark a backlash against conventional planning, a backlash that took Jacobs's *Death and Life* as a guide and users' manual.

Against the backdrop of New York City's struggles—symbolized by the loss of the Giants and Dodgers baseball teams to California in 1957 but painfully lived in white flight, labor strife, soaring crime, the blackout of 1965, and chronic financial woes culminating in near bankruptcy in 1975—analysts and observers mounted an urgent effort to extend Jacobs's insights to how cities actually worked and why the planning efforts were failing. This campaign was exemplified by architect and planner Oscar Newman's 1972 book *Design Guidelines for Creating Defensible Space*, which corroborated Jacobs's identification of public surveillance by "eyes on the street" and clear demarcation of public and private territory, creating a sense of ownership and responsibility, as critical to making safe places. A more positive effort involved celebrating the successes of the city through its diversity and grit in music, movies, and television shows like the wildly popular *Sesame Street*, which premiered in 1969 and was in a sense Jacobs's street "ballet" rendered with live people interacting on an "inner city" streetscape with fuzzy puppets of the baker, the policeman, the cook, and even the bum in the trashcan. Jacobs's effort was part of an upsurge of quasi-anthropological investigation of how cities worked by observing its denizens in action.

The condominium was one of the most significant developments in design and building of the twentieth century, and it is no coincidence that it appeared at virtually the same moment as Jacobs's book. The postwar boom in suburban single-family home building had largely left out the cities, as lending requirements made dense, multiunit projects difficult to finance, and renting lost its allure. Developers needed a different kind of property ownership to justify large projects for owner-occupiers. The first law legalizing separate ownership of units with joint ownership of common areas, the Horizontal Property Act, passed in Puerto Rico in 1958. Two years later, a Salt Lake City lawyer named Keith Romney (a cousin of Republican presidential candidate Mitt Romney), who had a client trying to develop a multiunit property, introduced similar legislation in Utah. The lawyer named the new form after an inscription he claimed to have seen carved into an ancient Roman ruin reading "condominio." Keith Romney became an evangelist for condo laws in other states, where developers and lenders eagerly embraced it as a way to combat "blight" without federally funded Urban Renewal by building taller and denser projects. In 1961, the Federal Housing Administration (FHA) began to insure condominium mortgages. The innovation was immediately popular in Florida and other places receiving an influx of Northerners fleeing cities and cold weather, and high-rises began to wall off the beach across the Gulf and Atlantic coasts. By 1969, every state had legalized condos. By 1970, there were 700,000 in the United States. Now, close to 30 million people—one in five Americans—live in one.

In their low-rise form, which as an architectural genre has taken its cues from Charles Moore's pioneering design at Sea Ranch, in Northern California, begun in 1965, condominium complexes are a perverse version of the garden city: miniature pseudo-villages clustered with their backs to the street, facing into bucolic interior common landscapes of paths, trees, lawns, and parklike amenities like fountains and benches, but all private, limited access, closed in on themselves. In their successful proliferation, condo complexes can be thought of as colonies, but unlike corals each one is unconnected to any other, nor to the fabric of the larger community around it.

Among architects, a reassessment of modernism gained momentum. Charles Moore was counted among the "postmodernists," who questioned modernism's faith in progress and rejection of architectural history and who sought to bring historical forms and references back into their practice. Robert Venturi, Denise Scott Brown, and Steven Izenour led a student studio to study the commercial forms of the Las Vegas Strip, baited functionalist modernism's "less is more" diktat (from Mies van der Rohe) with the quip "less is a bore," and urged a reconsideration of vernacular commercial architecture by stating that "Main Street is almost OK." Christopher Alexander, a British architect transplanted to Berkeley in the early 1960s, developed a "generative grammar" derived from traditional and vernacular buildings meant to be used by nonarchitects to design structures. The result, *A Pattern Language* (1977), grew to be one of the best-selling design books in history, followed by *The Timeless Way of Building* (1979), the two forming the backbone of a new, neotraditionalist movement with implications not just for designing buildings but for communities and cities.

In Europe, too, there was well-publicized interest in learning from, if not reviving, traditional architecture and urbanism, signally from the New Rationalists in the 1970s in France and Italy, led by Aldo Rossi and Rob Krier. Krier's younger brother Leon Krier, born in 1946 in Luxembourg, left his modernist studies in architecture at the University of Stuttgart in 1968 after only one year, moved to London, and worked with James Stirling until 1974, attracted by the British architect's questioning of modernist doctrine and inclusion of historical precedents in his buildings. Leon Krier practiced and taught for 20 years in London at the Architectural Association and the Royal College of Art, and has had many stints as a visiting professor at US universities including Princeton, Yale, and the University of Virginia. He became a fierce critic of functional zoning and of modernism's rejection of traditional limits of urban scale and size. Among his dicta is that city buildings ought to be limited to between two and five floors, but without limits on height—pointing out that the Eiffel Tower has just three floors—allowing for infinite variety but not for overly high densities. He published "The City Within the City" in 1977, proposing the idea of a coherent "European City" tradition "obtained from and verified by a millennia-old culture," which modernism had effectively crushed. His precepts are close to those of Jane Jacobs, with "quarter" replacing "neighborhood":

> A large or a small city can only be reorganized as a large or a small number of urban quarters; as a federation of autonomous quarters. Each quarter must have its own center, periphery and limit. Each quarter must be A CITY WITHIN A CITY. The quarter must integrate all daily functions of urban life (dwelling, working, leisure) within a territory dimensioned on the basis of the comfort of a walking person; not exceeding 35 hectares (80 acres) in surface and 15,000 inhabitants. Tiredness sets a natural limit to what a human being is prepared to walk daily and this limit has taught mankind all through history the size of rural or urban communities.

The city and its public spaces can only be built in the form of streets, squares, and quarters of familiar dimensions and character, based on the local tradition.

Like Jacobs, Krier laid blame at the feet of the architecture profession, his own: "After the crimes committed against the cities and landscapes of Europe over the last few decades in the name of progress and efficiency, the professions of architecture and engineering deserve nothing but the contempt of the population. The function of architecture is not, and never has been, to take one's breath away: it exists to create a built environment which is habitable, agreeable, beautiful, elegant and solid."

Over the years, Krier was tireless in expounding his ideas, and in the United States he gained at least two important converts: the architects Andres Duany and his wife, Elizabeth Plater-Zyberk. After graduating from Yale, the couple cofounded Arquitectonica, a Miami-based firm that gained quick renown for its colored, patterned, and sleek modernist buildings, including tall residential and office towers. Yet both were having their own doubts about the modernist program, and happening on Leon Krier's perspective was a trigger point. Duany explained:

> One day I went to a lecture by Leon Krier, [who] gave a powerful talk about traditional urbanism, and after a couple of weeks of real agony and crisis I realized I couldn't go on designing these fashionable tall buildings, which were fascinating visually, but didn't produce any healthy urban effect. They wouldn't affect society in a positive way. The prospect

of instead creating traditional communities where our plans could actually make someone's daily life better really excited me. Krier introduced me to the idea of looking at people first, and to the power of physical design to change the social life of a community. And so, in a year or so my wife and I left the firm and went off to do something very different.

Among Duany and Plater-Zyberk's first clients in their new firm, DPZ, was developer Robert Davis, who hired the architects to design a residential community on 80 acres on the remote Florida Panhandle coastline, land that Davis's grandfather had bought in 1946. Davis's inspiration came in part from the traditional beach towns in the area, with wooden vernacular architecture, porches, and metal roofs—definitely not the postmodern architecture Arquitectonica practiced. Duany and Plater-Zyberk decided to design the master plan for the community, to be called Seaside, and to leave the buildings to others, in order to allow architectural variety to arise, but within a carefully thought-through framework. They visited many traditional Florida towns and others across the South, measuring and observing how buildings interacted with the streets and how people moved through them. What they intended was more urban than a town and based on a couple of basic principles: there would be no functional separation of home, work, and shopping, and everything would be within a five-minute walk of a person's home. Accordingly, they planned for small lots and small setbacks from the street, and platted a network of pedestrian paths, many of them made of sand, organized with "vistas" of key buildings at the ends of major routes.

While working on a concurrent project, Charleston Place, in Boca Raton, Florida, the DPZ architects had found that many similar elements they wanted to include—parking lots and garages hidden in the rear of buildings, buildings built close to sidewalks, and narrow, tree-lined streets—were illegal under local zoning codes. For example, the streets they had designed weren't legally wide enough, and the buildings were too close to them. Duany and Plater-Zyberk were inspired by some of the surviving examples of small-scale urbanism in the country, but, they complained (in their 2000 book, *Suburban Nation,* written with Jeff Speck), "Somewhere along the way, traditional towns became a crime in America." Mostly, postwar planning codes had been written to accommodate car-oriented sprawl, with fast, wide streets and maximal amounts of parking. As a consequence, distinctive, small-scale urban places like Charleston, Boston's Beacon Hill, Nantucket, Santa Fe, Carmel, or Santa Barbara "exist in direct violation of current zoning ordinances," and certainly couldn't be built again. "Even the classic American main street . . . is illegal in most municipalities," they wrote.

Their solution at Charleston Place in Boca Raton was to label their streets as parking lots, because these weren't regulated as stringently, and treat their parking lots as streets. It was an imperfect solution to say the least, and at Seaside they resolved to write a new zoning code that would accommodate a compact urban vision oriented around walking, with cars in a subordinate role. The "solution" they wrote, "is not removing cars from the city," but to "tame" them through good urban design. They took their recipe for how to plan a neighborhood from an amalgam of the towns they had studied and small cities, best represented by Alexandria, Virginia, a city laid out at roughly the same time as Greenwich Village, and using similar principles, "following six fundamental rules that distinguish it from sprawl." Unsurprisingly, their recipe is consonant with that of Jacobs: One: the community has "a clear center," where the "civilized activity" of "commerce, culture, and governance" takes place. Two: all "the ordinary needs of daily life" lie within a five-minute walk from people's homes, mostly or completely eliminating the need to drive. Three: the community has a street network of small blocks—a "continuous web" making walking easy. Four: it has "narrow, versatile" streets, with wide sidewalks, parallel parking, and active storefronts, providing for slow car speeds and a rich pedestrian life. Five: it is mixed use, and yet is "not a design free for all," as types of buildings, and their implied uses, tend to cluster together—but it is the building type, not the use, that determines the clustering. Six: it sets aside "special sites for special buildings" such as churches and civic buildings "that represent the collective identity and aspirations of the community." Finally, like Jacobs, the architects assured that the sum of the rules would provide "the key to active street life . . . creating a 24-hour city"—letting cities compete with suburbs.

At Seaside, DPZ wrote an Urban Code based on the six Alexandria rules with specific conditions for the site. The challenge was to write a set of rules that would allow for variety, yet rule out what the architects considered incompatible buildings: "dingbat" apartment buildings built over street-front parking, for example, or buildings set back from the street in suburban style. It was coding to create diversity and spontaneity, within limits. Here was the seeming paradox of seeking the rules for self-organizing cities; or, in Jacobs's term, learning the rules of "organized complexity." The critical question was where to set the limits. For this, the designers used the concept of building "type," which they adapted from the ideas of the British architect Alan Colquhoun as well as Leon Krier, being characteristics of physical form such as height, massing, floors, and alignment to the street, sides, and rear, all of which strongly influence the best use for the building, but don't determine it. By guaranteeing "a consistent streetscape" the code "makes different uses compatible." Each area within Seaside was prescribed one or more of eight types, and each type was delineated in the simplest possible terms, more visually represented than in the numbers and words typical of municipal codes. In no case does type determine "style," which has more to do with materials, ornament, proportions, and angles. But as the developer Davis wanted a certain kind of vernacular-influenced architecture, DPZ wrote an additional architectural code, specifying construction materials (wood siding, doors, and windows; metal roofs) and some proportions (such as pitched roofs; and square or vertical windows, not horizontal) in order to achieve coherence in the buildings, but not sameness. Another example is the requirement that properties adjacent to the sand paths must have white, wooden fences, but not of the same style as another property on the same block. Outbuildings were specifically allowed, to encourage a range of flexible uses on a single property—aimed at promoting economic diversity, akin to Jacobs's "old buildings" principle.

A process for design review was provided, but any building proposal that meets the code criteria may go forward without further permitting issues. What DPZ hoped to achieve was "coherent urbanity," and at Seaside Duany and Plater-Zyberk succeeded on their terms: the community is a mix of styles, from Victorian to North Florida fish shack to neoclassical to muted postmodern, and a mix of forms, with varied kinds of viewing towers becoming a popular, but unforeseen, element. The intended qualities of "community" have in fact manifested in daily use: walking is more common than driving and the public gathering spaces are well used. Because no private walks over the dunes are allowed from the beachfront houses (their inhabitants must use the main walkovers everyone uses), the beach feels communal, with one result being that land values away from the beach have remained high relative to the beachfront—an unusual situation on the coast, where unbalanced land values have pushed developers to erect skyscrapers and slabs along the front row for hundreds of miles, blocking views and access.

The success of Seaside spawned more projects for DPZ, such as Kentlands, in Gaithersburg, Maryland (1988), in the suburban Washington, DC, area, a 352-acre development using colonial-style architecture, including row houses; and Windsor, Florida (1989), a 416-acre isolated resort development. Elsewhere, other architects and planners were working in a similar vein. In San Francisco, urban designer Peter Calthorpe published *Sustainable Cities* in 1986 with Sim van der Ryn and proposed the concept of "pedestrian pockets," walkable mixed-use neighborhoods linked by public transit to a larger region. In Los Angeles, Stefanos Polyzoides, a Greek architect who had come to LA in 1973 to teach at the University of Southern California, became enchanted by the courtyard apartment complexes common in the region in the 1920s and '30s, which were inspired by traditional Spanish buildings but adapted to early twentieth-century California: typically one-story rental cottages arrayed along a linear landscaped courtyard or path, affording privacy and greenery in a dense urban setting. Polyzoides undertook a careful study of the remaining examples, published as *Courtyard Housing* in Los Angeles in 1982, a pioneering examination of a single, though unusually capacious, housing form as a type. He met and married a colleague, Elizabeth Moule, and the two opened a Pasadena practice in 1990 devoted to architecture and urban design informed by the LA courtyards and Mediterranean tradition.

Nationwide, developers seeing growing market demand for alternatives to conventional suburban sprawl worked with these and other architects to supply it, with higher-density building types including row houses, with, at 18 to 24 units per acre, 5 to 10 times greater density than standard single-family houses, and duplex or triplex units, at 8 units per acre, double the density of single-family houses while looking like them. Combined with concerns about sprawl's environmental and public health impacts and high capital costs for infrastructure, these efforts formed part of what James Howard Kunstler, the author of *Geography of Nowhere* (1993), an antisuburban manifesto, called "a campaign to rescue the landscape, townscape and civic life of our nation from the failed experiment of a drive-in utopia." Continental Europe and the United Kingdom witnessed campaigns in similar directions. Much controversial impetus was provided by His Royal Highness Charles, the Prince of Wales, who aired his disapproval of modernist building and planning very publicly, using a speech to the Royal Institute of British Architects on its 150th anniversary in 1984 to denounce a planned tower addition to the National Gallery by Peter Ahrends as a "monstrous carbuncle on the face of a much-loved and elegant friend," and publishing a book in 1989, *A Vision of Britain*, with an accompanying BBC television documentary, detailing many more complaints. The prince's challenge helped stoke the formation of the Urban Villages Group in the late 1980s, which went on to have considerable influence on British government housing policy. In 1988, the prince hired Leon Krier to design a ground-up traditional community, Poundbury, on land he owns near Dorchester, in the county of Dorset, in the Southwest of England. The project, which broke ground in 1993, used a traditional street plan and a code enshrining similar principles to those at Seaside, and had several early buildings designed by Krier himself in traditional premodern European idioms. Poundbury predictably became a lightning rod for criticism: it was politically, no less than architecturally, reactionary, a medievalist fantasy out of touch with the modern world. At the opposite extreme, historical purists complained that the designs weren't confined to authentic, local Dorchester styles and materials, much less British ones, but were frivolously "romantic" like the eclecticism of the Victorian and Edwardian eras. Others rebuked it for elitism, being possible only in the completely controlled confines of the Prince of Wales's private Duchy of Cornwall, inside which Poundbury legally exists—even though affordable housing was included on the prince's express instructions. The most common complaint was that it's fake—a criticism also plentifully leveled at Seaside and Kentlands, though both Poundbury and Kentlands are functional communities, not resorts dominated by second homes, as at Seaside. Judging by surveys of residents' satisfaction and by property values two times higher than surrounding areas of the County of Dorset, the cavils haven't proved a problem. Other numbers, for example, those showing car use in Poundbury two times higher than in surrounding areas, would indicate that it might be a cloudy picture to render.

The same year that Poundbury began to take shape, Duany, Plater-Zyberk, Polyzoides, Moule, and the Bay Area planner Daniel Solomon together founded a new organization they dubbed the Congress for New Urbanism (CNU). They modeled the group on the modernists' CIAM of 1928, including adopting a charter of basic principles, as CIAM had with its Athens Charter—although the CNU group understood that the precedent was ambiguous, since CIAM, in Duany's phrase, "can be credited or blamed" for the direction that city building took after Le Corbusier effectively took it over, the very direction that CNU was formed to combat. The Congress dedicated itself to combatting suburban sprawl and to the promotion of traditional neighborhood development through a list of the principles that Jane Jacobs had enumerated and that others, including Krier and CNU's members, were continuing to expand and delineate. To the ideas of walkability, mixed economic use, active street life, and high-quality design, they added the concept of the "transect": a hypothetical line drawn through any given settlement placing the highest density in the center or city core, and gradually decreasing toward the periphery, finally becoming farms, then nature. In sum, these elements would "add up to a high quality of life well worth living, and create places that enrich, uplift, and inspire the human spirit." The last principle seemed more like a hope, similar to, but several philosophical steps beyond, Jacobs's commitment to giving "city life . . . its best chances." What had not changed between her *Death and Life* and the Charter for New Urbanism was the focus on the mixed, active street as the elementary particle of good urban design. With its members as evangelists for the cause, speaking, writing, designing projects, and leading "charettes," as they liked to call the community-inclusive design workshops often held to achieve buy-in from officials and neighbors, the Congress grew quickly and drew press attention for its ideas as much for its built experiments like Seaside. Andres Duany in particular, with groomed good looks and an articulate, convincing delivery, has been a ubiquitous presence in lecture halls and in the press. CNU became international, counting 3,000 members, holding packed annual conferences, and gaining influence on real-world policy, from national directives to city and town codes: New Urbanist principles influenced the US Department of Housing and Urban Development's HOPE VI guidelines, as well as the Urban Land Institute's handbooks for developers.

In 1994, just a year after Poundbury did, another New Urbanist community broke ground: Celebration, in central Florida outside

Orlando, on land owned by the Walt Disney Company adjacent to its Disney World and Epcot resorts. On the basis of extensive market research, the company chose New Urbanist design principles for a town to eventually house 20,000 residents and hired the architect Robert A. M. Stern, an early postmodernist and later neo-neoclassicist, to master plan it. Stern drew up a town plan based on mixed-use neighborhood centers, pedestrianism, and "community" values. Six housing styles would be offered: Classical (a version of the Greek Revival found in antebellum Southern towns like Natchez, Mississippi), Victorian, American Colonial Revival, Coastal (like Seaside's updated fish shacks), Mediterranean, and French (really nodding to New Orleans, not France). The first phase opened to enormous excitement, market interest, and press attention, in no small part because it was the first time a major corporation had put its name on a master-planned community with so prominent a utopian mission—it must be called that, even though Celebration's was a very mild form of Utopia, advertised as a kind of paradise where one's problems would be solved through design. (It is worth noting that the Disney Company's early working name for the project was Dream City.) Celebration sold, and rented—there were apartments in the town center—briskly. Some issues emerged: Disney's land was a long distance from the vast majority of work and shopping opportunities in the area, so car use was high—and the problem was exacerbated by high home prices (the median home price in Celebration was two times higher than in the greater Orlando region in 1997), which, in forcing so many spouses to join the workforce to cover mortgage costs, led to more drivers on the road. The prominent little Eden also contrasted sorely with the surrounding blue-collar area along State Route 192, with its strip malls, honky-tonks, and cheap entertainment centers—a perceived social gap widened even more by the town's designer Stern blurting out that Route 192 was "the sleaze road of all times."

As had been the case with its sister communities, Celebration drew accusations of nostalgia for a vanished or outright mythical small-town America, and worse, for using historical quotation as a marketing device. This perception wasn't helped by the fact that it was built by Disney, a corporation globally associated with the total control of people's experience in artificial environments. Was Celebration a real place or an expensive theme park for adults to play house in? The question became a national topic of some amusement with the 1988 release of the movie *The Truman Show*, starring Jim Carrey as a man who since childhood has been unknowingly living inside a fake town, with everyone around him actors deluding him into believing the fakery, and his every movement, including sleeping, recorded by hidden cameras and broadcast 24 hours a day as the world's most popular TV show. The movie as it happened was filmed at Seaside, not far away, but it was Celebration that was typecast as a stage set, and New Urbanism along with it as a suspect, inauthentic phenomenon rather than a return to an authentic, and better, past.

Many in the architecture establishment dismiss New Urbanism as an exercise in retrograde historical revival. Proponents and practitioners respond that style is not the point; form is. Duany has written: "We are prepared to sacrifice architecture on the altar of urbanism . . . because architecture is meaningless in the absence of good urban design." In any event, he and others reject the idea that New Urbanism's typical architectural choices are historicist, preferring the label "neotraditionalism," which they say is "nonideological" and defined not by slavish adherence to precedent but by picking and choosing among old and new elements, as in, for example, "the Mazda Miata, a car that looks, sounds, and handles like a British roadster but maintains the rate-of-repair record of a Honda Civic." Duany and other New Urbanists also use the defense that traditional style is a market necessity: "It is hard enough convincing suburbanites to accept mixed uses, varied-income housing, and public transit without throwing flat roofs and corrugated metal siding into the equation." Thus, in John Dutton's words, neotraditional style is used "as camouflage for subversive density, difference, and mixed use." In an interview, Leon Krier acknowledged that "the situation is so critical that Andres Duany and I have discussed for a while designing a modernist town simply to show them how it is done. A town design code could easily limit itself to Le Corbusier's 1920s or 1950s grammar and produce a meaningful townscape; the same could be done with Frank Lloyd Wright—or even Zaha Hahid or Oscar Niemeyer idioms."

In the meantime, New Urbanism has proven itself in the commercial marketplace, and developers increasingly turn to it, or a version of it, for a range of projects—including a growing number of examples of conversions of failed suburban shopping malls into mixed-use communities. It and its offshoots and allies form the most committed, evolved, and assertive parts of a broad front bent on reinhabiting central cities—what Robert Fishman has dubbed the "fifth migration"—the dominant phenomenon in urban life in the early twenty-first century. But with success comes the danger of being watered down: examples abound of cheap rip-offs that call themselves New Urbanism but are really no more than conventional suburban tracts, malls, or apartment buildings trussed up in architecturally traditional or at minimum diverse dress. Even for legitimate developments, the problem of gentrification hangs over the effort, especially with infill or conversion projects in cities. Jane Jacobs was widely accused of promoting gentrification with her selling of older neighborhoods like Greenwich Village to what was de facto an educated middle-class audience. It is no coincidence that her former home at 555 Hudson Street, a 2,144-square-foot building in the heart of what has become one of the most trendy districts in the United States, sold in 2009 for $3.3 million. As the movement continues to gain influence both on policy and in the built environment with its codes, which were meant, in the words of the 1920s City Beautiful planner John Nolen, to be "safeguards against incongruity," it risks ossifying into something rigid that stifles creativity and diversity or serves gods other than those virtuous ones worshipped by the New Urbanists. In another irony, New Urbanism's real-life elevation of the figure of the enlightened architect and planner saving us from bad urban design through "good architecture" echoes the championing of the county architect by Frank Lloyd Wright and of the all-powerful enlightened despot by Le Corbusier. Codes, after all, are like any form of power: they can be used for good or for ill. It's worth remembering that the first codes that completely transformed Los Angeles and New York in the first two decades of the twentieth century promoted public health and efficiency but also ruthlessly excluded some groups from those fruits and distorted markets in favor of empowered classes. Coding corals, even if it isn't an outright paradox, presents the enduring challenge of balancing order versus freedom—a tightrope act, and there is understandably much skepticism from the gallery. Perhaps the best summary—and warning—is provided by the character of Christof, *The Truman Show*'s creator and producer in the movie, played by Ed Harris, who whines in defense of the world he's made and managed so successfully for so long: "It's not fake. It's just controlled."

5. Corals

Field Guide: New Urbanism

Diagnostics

- Code-controlled architecture: groups of buildings or entire communities share a single historicist style, or a limited range of historicist styles; less often a disparate range of styles including modern variants. Building form is constrained to a range of heights and sizes, grouped to be limited to specific streets or neighborhoods. Signage, street furniture, trees, and landscaping are designed according to a planning code.
- Variety: within limits prescribed by code, the style, colors, and materials vary; building heights, masses, and setbacks from street vary.
- Neighborhood planning is based on traditional walkable neighborhood, with central street(s) for shopping and services.
- Street grid is as dense as possible; not interrupted by superblocks; pedestrian streets and connections between streets, parking, and buildings are created where possible.
- Mixed use: neighborhoods consist of deliberately mixed residences, retail and other businesses, places of employment, schools, and recreation.
- Multimodal transportation: neighborhoods are pedestrian-friendly; bike-friendly; accommodating cars in a nondominant mode with narrow streets, on-street and off-street parking; ideally incorporating mass transit options such as bus, trolley, and light rail.

Examples:

- Florida: Seaside; Celebration; Charleston Place, Boca Raton.
- Kentlands, Maryland.
- Poundbury, Dorchester, England.
- Jakriborg, Sweden.

Variants:

- Converted malls: Belmar, Lakewood, Colorado (former Villa Italia mall); CityCenter, Eagleton, Colorado.

- Pseudo–New Urbanist Infill: these areas mimic the architectural appearance of New Urbanism but are car-dependent, without a complete street grid, or surrounded by fast roads and parking lots.
- Overlaps common: China in chapter 1; shopping malls with residences mixed in, chapter 6.

Images:

Seaside, Florida (1980s). Plan: DPZ.

Celebration, Florida (1996–). Plan: Cooper, Robertson and Partners and Robert A. M. Stern. ***Photo taken by Bobak Ha'Eri. February 23, 2006***

"The Whistling Witch," Poundbury (1993–), Dorset, England. Plan: Leon Krier.

Boulevard du Lac, in the Beverly Hills private townhouse development, Tai Po, Hong Kong.

Jakriborg, (late 1990s–), Sweden, New Urbanist–style new town. Architects: Robin Manger and Marcus Axelsson.

Chapter [6]

Malls

Victor Gruen, Jon Jerde, and the Shopping City

Not only is shopping melting into everything, but everything is melting into shopping. Through successive waves of expansion—each more extensive and pervasive than the previous—shopping has methodically encroached on a widening spectrum of territories so that it is now, arguably, the defining activity of public life.

—Sze Tsung Leong, *Harvard Design School Guide to Shopping*

I can't remember my first visit to a big shopping mall—it seemed always to have existed as the place one's family drove to park the car, then walk around, dipping into shops, traversing a department store, and sitting down for something to eat. It seemed both an indispensable destination for accomplishing certain errands and a vague place to pass the time and wander without too much forethought or intention. But I do remember the first time I became aware that the mall was an important—and in some way controversial—part of the cultural landscape I lived in. It was the first time I heard on the radio Frank Zappa's 1982 song "Valley Girl," about the lifestyle of teenage girls in an upscale part of the San Fernando Valley in LA, played out in a brand-new mall—the Sherman Oaks Galleria on Ventura Boulevard, which had opened to much fanfare in 1980.

Zappa had enlisted his 14-year-old daughter Moon as the singer, showcasing her impression of conversations overhead between "Valley girls" in their signature "Valspeak," full of "like" and "totally" and "Oh my God!" squeaked out with a prophetic upturn at the end of the phrase. The intent was a satirical, even acid, critique of Southern California suburban youth culture as vapid and consumerist. But, in perfect revenge on the ironist Zappa, the song became a surprise radio hit, rising to number 32 on the pop charts—Zappa's only US Top 40 ranking. The same year, the comedy film *Fast Times at Ridgemont High*, starring Sean Penn as a stoned surfer, then the 1983 film *Valley Girl*, both shot at the Sherman Oaks Galleria, shined even brighter light on the mall as a crucial social space for young people, where the newest, coolest things could be found, including new ways of speaking and dressing. Both song and films were satires, but they successfully propelled mall subculture to mass awareness, and gradual adoption, helping to raise this newest yoking of commercialism, youth culture, and architecture to the status of uncontested global idiom and standard. Shopping, they said, may be risible, but it is also irresistible—because it promises not just new shoes or clothes, but the possibility of a new identity.

Over the years a progression of malls in the Los Angeles area, well publicized locally and nationally, charted a ratcheting evolution of the form from the standard suburban, enclosed mall with double department store anchors toward something new. The Galleria was joined in 1980 by Frank Gehry's Santa Monica Place, then in 1982 by the Beverly Center, with its biggest-ever 14-screen multiplex, and original Hard Rock Cafe (the mall was the set for the 1991 film *Scenes from a Mall*, starring Woody Allen); all were essentially still the standard suburban model, imported into the suburb-cities of Santa Monica and Beverly Hills. But a slow responsiveness to the city had begun to manifest itself. Jon Jerde's remodels of the Westside Pavilion in 1985 and Fashion Island in Newport Beach in 1988 upped the ante with a gradual opening-out of the introverted mall to the sky and streets outside. Then, in 1993, came the éclat of Jerde's CityWalk at Universal Studios, a pedestrian "street" leading from the parking garages to the entrance of the theme park, staged to look like a madcap, carnivalesque distillation of LA urbanism, with storefronts mixed with cafés and leaping fountains, bright colors, and over-the-top signage. Aside from a huge overhead trellis in its central area, it had no roof—the indoor mall had broken outside, while still being behind the doors of a private, enclosed development. The trajectory steepened: soon, more malls tried on the "city" look with historically themed pastiche architecture, streetcars on tracks running down the middle of the "street," and a panoply of constant entertainment in the form of music and performers—and real inhabitants. The Grove in Hollywood (2002) and the Americana in Glendale (2008), both by developer Rick Caruso, touted luxury residences that topped retail streetscapes and were serviced by private elevators, rooftop swimming pools, and valet parking. These were perfect consumer planetoids, undeniably informed by Disneyland and Las Vegas, but they were no longer somewhere you went, but somewhere you aspired to live. And these, it turned out, were small potatoes: in the past decade and a half, entire districts of cities have been configured as full-service, live-in malls: in Dubai, Singapore, Shanghai, Seoul, and many, many other places. Now entire cities designed and built from scratch are extending the reality of shopping cities to larger and larger expanses of the globe, from Russia to Indonesia to China and back to Southern California.

If the world is becoming a mall, has shopping become the driver of urban form? In most educated circles this suggestion elicits a collective shudder. Shopping is sub-serious, as Cicero insisted: "All retail dealing may be described as dishonest and base." Architecture, always zealous in defense of its claim to be a high art, wants nothing to do with it. Except, on rare occasions, to pay the bills. Louis Sullivan did a department store, Frank Lloyd Wright a boutique, Rudolf Schindler a store or two, and I. M. Pei's first major project was a mall, but these are rarely mentioned along with their canonical "masterpieces." And yet a case can be made that shopping, in the form of trade, gave birth to the city, that shopping has been and remains the lifeblood coursing through its heart, that the design of shopping is inseparable from the design of cities since time immemorial and is an indispensable guide to the urban future.

The largest neolithic settlement known, Çatalhöyük in Turkey, was founded in 7000 BCE, probably as a trading center. The market at the center of Thebes has been dated to 1500 BCE. The Greek agora, or "gathering place," the acknowledged birthplace of Western civilization and democratic society, was both a marketplace for shopping and a civic center for discussion, sociality, and politics. The Greek words for "I shop" and "I speak in public" are both derived from the same root; in modern Greek *agora* still means marketplace. The agora became the Roman forum, the medieval fair and market town, the Eastern bazaar and souk. Is shopping a recipe for the city? Consider the evidence. In the exchange of goods is gathering, and in gathering is society: meeting, trading information, gossiping, haggling, freedom of movement for women, and people-watching—the original theater is the theater of customers as participants in a perennial ritual and unpredictable drama. Done right, shopping can define space in ways that are fundamentally urban: the shopping space is a space apart, inside, separate from other distracting activities, and essentially pedestrian, but also connected to the outside.

Shopping generates movement and density; it mixes and connects people, and disconnected or disparate parts of the city. If this is the case, then maximizing shopping equals maximizing urbanism.

The historical record points to continual innovation in the physical forms of shopping, which in turn changed the social dynamics of the culture. In the seventh century BCE permanent retail shops appeared in Lydia, Greece. In Rome, making space for shopping was regarded as basic to the civic order: in 45 BCE, as one of his last policy innovations, Julius Caesar banned carts from sunrise to sunset; sidewalks were ubiquitous, an emblem of Roman cities. With the empire's fall in the fifth century, sidewalks went into decline, and so did shopping, until growing commerce and retail trade returned early in the second millennium, triggering the great urban growth of the Renaissance. Rome also pioneered the covered market, roofed from weather, and separated from outside, with Trajan's Market, in about the year 110, where stone arches formed two symmetrical arcades under a vaulted roof. In 1461, the same idea reappeared in the eastern Mediterranean with the Great Bazaar in Istanbul, then in Europe—notable examples being the Royal Exchange, in London in 1566–68; followed in 1606 by the New Exchange in London; the 1608 Amsterdam Exchange; and the 1667–71 Second Royal Exchange in London.

These were spaces of commerce, but not strictly speaking spaces open to the public. The big breakthrough came in seventeenth-century Paris, at the Palais-Royal, a neoclassical pile built around a large garden facing the Louvre across the Rue Saint-Honoré and the Rue de Rivoli, near the Tuileries gardens and the center of royal and fashionable Paris. Originally built for Cardinal Richelieu in 1634, the palace was given by Louis XIV to his brother, the Duke of Orléans; it assumed a long career as a focus of high-society gatherings, with a changing array of residents, guests, and celebrated incidents. In

1784, the then duke, Louis Philippe II, looking for sources of income, opened the renovated palace and grounds as a shopping and entertainment complex, with shops, salons, cafés, galleries, theaters, bookshops, bars, and brothels to stop in, and courtyards, avenues, fountains, and gardens to stroll in. It was the world's first integrated mall. Theaters stood at either end—the great playwright Molière staged his works there in the seventeenth century; the Comédie-Française has called it home since the eighteenth. For other tastes, prostitutes strolled the arcades, and gambling took place on the second floor. The duke allowed everyone in, regardless of class, and there was something for everyone: it was at once marketplace, social promenade, and political space. But its biggest audience was the new, burgeoning bourgeoisie, or third estate, the class created by and driving forward the consumer capitalism that was reshaping the world. During the revolution, the duke contracted revolutionary fervor, changing his identity from Louis Philippe II, Duke of Orléans to Philippe Égalité, and the Palais remained as popular as ever. It was a place for café society, with its politics (the Marquis de Sade in *Philosophy in the Bedroom* noted that good feuilletons, or political pamphlets, could be got there) and intrigue (it was a hotbed of Freemasonry, among other political persuasions), alongside a range of entertainments from the ladies of the night to the cannon in the garden fired every day at noon, its fuse lit by a sundial equipped with a magnifying lens. A poem by Jacques Delille captured the atmosphere:

Dans ce jardin on ne rencontre,
Ni champs, ni prés, ni bois, ni fleurs.
Et si l'on y dérègle ses mœurs,
Au moins on y règle sa montre.

(In this garden one encounters neither fields nor meadows nor woods nor flowers. And, if one upsets one's morality, at least one may reset one's watch.)

In 1786, the duke invested in stone colonnades enclosing three sides of the garden, calling them the Galeries des Bois, after these nonexistent woods. They were filled with more shops and cafés, drawing more customers into and around the enlarged space. He had created the first arcade, an innovation that simultaneously achieved several things: it provided an accessible, pleasant public space within private property where people could mingle, circulate, and window-shop, without the distractions and danger of the traffic, noise, filth, and smells of the sidewalk-less Paris streets; it offered new marketing possibilities for luxury goods and services; and a model for real estate speculation in properties being made available in the central city by government expropriation and expulsion of the poor. As a form, it was similar to the shop-lined covered bridges popular in many cities, such as the Ponte di Rialto in Venice, the Ponte Vecchio in Florence, London Bridge, and the Pont Notre-Dame, Pont au Change, Pont Marie, and Pont Saint-Michel, nearby along the Seine in Paris. But it also clearly drew stylistically from fashionable gardens, where people might stroll, mix, and enjoy the beauty and rarity of fine things. Crucially, it allowed women a social freedom impossible either in the isolation of the home or in the chaos and menace of the city streets.

In its next iteration, the arcade took its key technical ingredient from gardens: the glass-and-iron shell, borrowed directly from the glasshouses growing ever larger on aristocratic estates and public gardens all over Europe, providing lighting and protection from the weather, and allowing property owners to quickly and cheaply roof over passageways between buildings without costly foundations. In 1791, the Passage Feydeau set the essential form: a pedestrian passage connecting two streets through the middle of a block, lined with symmetrical, facing stores, closed off with glass fronts, and roofed with glass skylighting. It was a new kind of street: providing pedestrians a protected, separate, limited-access milieu; landowners new investment opportunities; and merchants new selling opportunities. The concept quickly spread: the Passage Feydeau was followed in 1799 by the Passage du Caire (named for Cairo, where Napoleon was fighting to break the British monopoly on trade in the Mediterranean), complete with an Egyptian decorative motif, and in 1800 by the Passage des Panoramas. Then followed a cascade of arcades: 15 were built between 1820 and 1840, making central Paris into an city of arcades, where it was possible to walk through much of the center inside a continuous, shop-lined, profitable, middle-class realm. An illustrated guidebook explained the innovation in 1852: "These arcades, a recent invention of industrial luxury, are glass-roofed, marble-walled passages cut through whole blocks of houses, whose owners have combined in this speculation. On either side of the passages, which draw their light from above, run the most elegant shops, so that an arcade of this kind is a city, indeed, a world in miniature."

The arcade jumped the English Channel in 1815–19, with the Burlington Arcade, off Piccadilly in London, then spread to other cities becoming wealthy with trade: Bordeaux, Brussels, Glasgow, Newcastle, even Saint Petersburg, where the Passazh opened in 1848. The height, size, and complexity of the new arcades pushed ever upward, reaching a peak in Milan's 1865–67 Galleria Vittorio Emanuele II, with its double arcade covering four levels, intersecting in a towering glass dome. By 1890, the United States could boast the largest arcade in the world, Cleveland's 300-foot glass dome covering five stories of shops and offices. At the same time, a campaign of *trottoir*, or sidewalk building, extended the success of the arcades out into the streets: between 1838 and 1870 more sidewalks were laid in

Paris than at any time since imperial Rome, totaling 181,000 meters in 1849, of a total of 420,000 meters of streets.

Walter Benjamin, the great literary critic and social historian of the early twentieth century, saw the Paris arcades as the crucible of nineteenth-century society, especially during the conservative Second Empire of 1852–70, when industrial production, commerce, consumer capitalism, and a repressive bourgeois social order came together to create modernity. It was fitting to him that the display of clothing was at the center of the spectacle, as the textile trade on one hand and the industrial revolution with its iron and glass on the other provided both the materials and the wealth for the entire system. Benjamin quoted the novelist Honoré de Balzac to describe "the time" of the arcades as one of retailing clothing: "The great poem of display chants its many-colored strophes from the Madeleine to the Porte-Saint-Denis." Benjamin noted further that, to extend the shopping hours into the night, the arcades were "the scene of the first gas lighting." These new spaces and their activities gave rise to a new kind of man: the flaneur, immortalized by the poet Baudelaire, who strolls, window-shopping—not for goods but for new kinds of things, people, and social arrangements, who "leisurely . . . goes botanizing on the asphalt," giving the arcade-city "its chronicler and its philosopher." For Benjamin, this was the first modern literature, conceived through Balzac's and Baudelaire's struggles with the new urban space of constant movement—the constant movement of shopping.

Benjamin saw in the arcades the apotheosis of the new form of fashionable commodity that drove capitalism's spread, what Marx memorably called "commodity fetishism," where relations between persons could be displaced into objects, casting them in a quasi-magical role: "The result of this displacement is the point where social relations enter the realm of the fantastic. The fantastic arises as the fetishized object takes on a value unrelated to its material existence. The object subsequently takes on a life of its own." Freud, likewise, saw fetishization in the new world of fashionable commodities: human sexual desires could also be displaced onto objects. For Benjamin, the two insights combined explained how the transaction of the fetish-commodity, which took place preeminently in the arcades, was the drug inducing "the new dream-filled sleep" that "came over Europe," a return of the mystical thought that modernism claimed to have overcome. He clearly understood that this transformation wasn't possible without the stage set of the arcade: the iron-and-glass "fairyland" where capitalism connected to the world of dreams.

On the heels of the arcade came another new kind of shopping architecture, also roofed in iron and glass, growing from the tradition of the exchange building or bazaar: the gallery. Where the arcade connects passageways by being a kind of street, the gallery connects one building to another with a transitional space like a plaza, but fully enclosed and roofed, lit from above by a light well or atrium, and lined with shops, often in multistory stacks. Galleries or bazaars were commonly built in series, like linked courtyards or rooms—again taking cues from the layout of formal gardens. From 1816 to 1840 many were built, especially in London, with 15, versus 4 in Paris and 1 in Manchester. Then the gallery broke free: in 1851, as the centerpiece of London's Great Exhibition, in Hyde Park, Joseph Paxton erected the Crystal Palace, an enormous glass building 1,848 feet long, 408 feet wide, and 108 feet high, made with 900,000 square feet of glass. It was really two series of galleries, alternating between 24 feet wide and 48 feet wide, flanking a central nave 72 feet wide. It easily enclosed six existing elm trees in the park, instantly making it feel like a landscape garden. Paths winding between parterres, statues, and fountains, including a 27-foot-high Crystal Fountain in the central crossing, only solidified the effect. The interior covered 19 acres—four times the size of Saint Peter's Basilica in Rome, and six times the size of Saint Paul's Cathedral in London. Inside were 100,000 carefully arranged exhibits featuring products, goods, and machines, displayed by nearly 14,000 exhibitors.

This first mega-mall came directly from the garden: Paxton, born in 1803, the seventh son of a Bedfordshire farm family, became a garden boy at 15, then moved up to the Horticultural Society's gardens at Chiswick, where he impressed the Duke of Devonshire, who offered him, at the age of 20, the job of head gardener at his grand estate, Chatsworth. There Paxton experimented building fountains, forcing houses, and glasshouses, culminating in the gargantuan Great Stove or Great Conservatory in 1837: 227 feet long and 123 feet wide, roofed with a rib-and-furrow system of radiating thin beams supporting plate glass, with heat provided by a boiler distributing steam through iron pipes. In 1849, Paxton managed to get the exotic Queen Victoria water lily, recently sent back from the Amazon, to flower, when the royal gardeners at Kew had proved unable. Watching it lifted out of its too-small 12-foot tank, he noted the leaf's structural system of ribs and webbing—which gave him the idea for his extraordinary glass palace. He sketched the initial concept in June 1850, and the building was ready by May of the next year, hosting 6 million visitors from all over the world during its six-month run.

Critics of the Crystal Palace gave it grudging admiration: "We freely admit, that we are lost in admiration at the unprecedented internal effects of such a structure . . . a general lightness and fairy like brilliance never before dreamt of, and above all . . . an apparent truthfulness and reality of construction beyond all praise. Still the conviction has grown upon us, that it is not architecture; it is engineering—of the highest merit and excellence—but not architecture."

The building's soaring scale shocked, but its light, and its seeming dissolution of the line between interior and exterior space, caused true wonder. One visitor wrote that Paxton had turned air into its solid form. A German writer exclaimed: "The total effect is magical, I had almost said intoxicating. The incessant and never-ending motley of forms and colors, the transparency on every side, the hum and buzzing in every direction, the splashing water of the fountains, and the heavy measured beat and whirl of machinery, all combine to form a spectacle such as the world will scarcely behold again."

These breathtaking new structures of the Victorian heyday, fairylike and magical, inspired an outpouring of creativity. Some visionaries proposed Utopias, like Fourier's phalansteries, enormous communal housing complexes, directly influenced by the Palais-Royal. In the wake of Paxton's achievement, schemes surfaced for large-scale arcades as urban transportation systems, including William Moseley's Crystal Way of 1855, a circulation system of 3.8 kilometers of underground trains with a pedestrian arcade above it lined with 5.3 kilometers of shops, to link London's West End and the City. Not to be outdone, the same year Paxton proposed the Great Victorian Way: 16 kilometers of continuous arcades to run between every London train station, thereby eliminating street traffic. Though these utopian schemes remained paper dreams, more practical structures were built in short order: exhibition halls—another Crystal Palace Dublin in 1852–53, and the Industrial Exhibition in Paris in 1855, market halls such as the Halles Centrales in Paris in 1863, library stacks, and prisons. But the greatest application of the new forms was undoubtedly the great Victorian railroad stations—which are in essence large arcades, connecting different cities, not just different streets, with great spaces of transition and movement—and shopping. Notable examples were Newcastle Central station, built in 1846–50; London's Paddington station, built in 1852–54; and King's Cross station, built in 1850–52—huge, glass-roofed sheds containing within themselves nearly the entire panoply of wonders as the Palais-Royal had two generations before. (The railroad station would in turn become the airport.)

The lessons of the arcade and the gallery were not lost on retailers. The department store, which had been gestating since the late eighteenth century in France, was given a big lift by the Crystal Palace and its new kind of place-market-spectacle that intoxicated visitors. Harrods was established on Brompton Road, Knightsbridge, just south of Hyde Park, in 1849, to capitalize on the coming Great Exhibition, and swiftly incorporated the lessons learned in a soaring interior space fitted out with special displays, lighting, and attractions. The Bon Marché in Paris, expanded in 1854, did the same, followed in New York by Macy's, opened in 1858, and Bloomingdale's, opened in 1872. Every major city had a department store before long. Glass made possible soaring atrium spaces that were routinely compared to cathedrals. Émile Zola, in his 1883 novel *Au Bonheur des Dames*, called the department store "a cathedral of modern trade, light yet solid, designed for a congregation of lady customers." The architectural innovations were matched by ever-increasing selection, low prices, customer service in the form of free delivery and returns, food establishments, and entertainment—performances, concerts, and art exhibitions became commonplace. Department stores developed into not merely places to shop, but popular destinations in their own right. Even socialists loved them: Ebenezer Howard planned his garden city around an enormous emporium called the Crystal Palace—putting the mall at the literal and figurative center of Utopia.

In the United States, it is no surprise that the model got a big bump around the turn of the century, in Chicago, at the hands of Daniel Burnham, deploying the same steel frames, elevators, and glass that led to his skyscrapers to design ever-larger and more spectacular stores for his merchant clientele. Burnham had already been building light-filled, soaring spaces with iron-and-glass greenhouse ceilings and intricate iron staircases connecting ground-floor public areas to balconies: the Railway Exchange and the Rookery in Chicago, built in 1885–88, and the Ellicott Square building in Buffalo, built in 1894–96. Amplified in his department stores, these spaces became instant landmarks: Marshall Field's, built in Chicago in 1902–7—with its Louis Comfort Tiffany blue glass mosaic ceiling of the south atrium, the largest in the world, and its 13-story north atrium—was the largest store in the world for a stint, but was eclipsed by the store Burnham built for John Wanamaker in Philadelphia in 1909, which encompassed 1.8 million square feet and covered an entire 480-by-250-foot city block. Burnham himself boasted that it was "the most monumental commercial structure ever constructed anywhere in the world. Its total cost has exceeded Ten Million Dollars." President William Howard Taft spoke at the opening dedication, making his entrance at the head of a brass band and followed by generals and admirals in dress uniform. Burnham's firm followed these successes with another Wanamaker's in New York, Filene's in Boston, Gimbels in New York and Milwaukee, May Company in

Cleveland, and Selfridges in Oxford Street, London, commissioned by Gordon Selfridge, who had worked for Marshall Field's in Chicago, and was the first major use of steel frame construction in Britain. Over the course of his career, Burnham, far better known for his skyscrapers and city plans, built 14.7 million square feet of shopping buildings, making up nearly a third of his total output.

Innovations traveled far, and fast. As part of its westernizing zeal, Japan began building railroads in earnest around the turn of the century, and then, as in a natural progression, department stores, called *depatos*. The first was the Mitsui kimono store, in 1900. Like Paris's Bon Marché or Wanamaker's, depatos had music halls and art galleries, from at least 1910 on. Most depatos were built by the railroad companies at the sites of their terminals and stations, connecting their trains to the urban subway system and directly integrating shopping into the city's mobility and infrastructure. This pattern holds today, as major retail complexes are built around public transport stations, forming the nuclei of major districts. In Tokyo, the areas of Shibuya, Shinjuku, and Ginza—fundamental parts of the warp and weft of the city—are shopping mall districts centered on train stations.

Constant change was the norm: electricity allowed interior lighting, obviating the need for windows; elevators were joined by escalators, connecting floors within the great atriums, allowing more horizontal layouts; cooling pipes and then air-conditioning further severed any necessary relationship between inside and outside; and competition from other forms of entertainment, especially movie theaters, which themselves incorporated many of the architectural tricks of the retail business, forced stores to keep innovating. But the biggest change came into view just as department stores reached the pinnacle of their size, complexity, and market domination:

decentralization, begun by streetcars and commuter railroads, and massively accelerated by the automobile. A growing middle class, attracted by cheap land, expanding regional road networks, and affordable car ownership, moved to the periphery, and, though middle-class consumers still traveled downtown to shop, they were increasingly frustrated by the congestion and parking shortages that plagued central cities. Entrepreneurs responded by building service and retail developments in outlying districts. The long process of breaking down the dominance of downtown began in the 1910s with gas filling stations. Gradually, owners added related items and services to the gas pumps, becoming superstations. It was a short step to the drive-in market offering groceries and dry goods, with a forecourt for parking. The first appeared in the Los Angeles area and quickly became ubiquitous there. In 1928–29, Chapman Market set the standard for the upscale neighborhood shopping center: with Mayan Revival styling, a mix of shops and restaurants, and off-street parking. Modernist architect Richard Neutra scaled the concept up with his proposed 1929 Dixie drive-in, a supermarket facing a large parking lot, and the enormous possibilities of the form became apparent. By the early 1930s, the drive-in supermarket was standard in the LA region and became the national model in the 1940s. The buildings' exteriors began to matter less, while parking and circulation to accommodate motorists mattered more, and took up more and more space. Buildings flattened to one story. Store interiors grew less hierarchical, divided into many identical aisles and geared toward self-service, allowing shoppers to navigate at their own pace and direction, with less or no contact with sales staff, and a single point of payment at exit registers.

Even as new shopping forms appeared and evolved, older ones, once seen as the utmost vanguard of modernity, in their turn became obsolete. In the 1920s, the Paris arcades, having fallen into decline with the rise of the department stores, were photographed as found-art curiosities by artists such as Germaine Krull and Eugène Atget, and their images of dilapidated shop-window mannequins and dummies, some with their heads detached, were celebrated by the Surrealists for their uncanny evocation of liminal psychological states.

The future of retail became visible in 1922, in a rural area dotted with pig farms four miles south of downtown Kansas City, Missouri, incongruously clothed in the architectural imagery of baroque Spain. There, developer J. C. Nichols laid out an upscale residential development of themed historic homes he dubbed the Country Club District, arrayed around a carefully planned package of amenities: Ward Parkway, a wide landscaped boulevard ornamented with statues and fountains to whisk motorists to and from downtown, the eponymous golf club, and a shopping center—Country Club Plaza—a village-like cluster of shops, services, and restaurants, with ample parking, themed in the Spanish Colonial Revival. One writer called the plaza "Mediterranean Kansas," as part of the Country Club District was in neighboring, and more desirable, Kansas City, Kansas. Nichols intended the shopping center to function as a kind of town center, substituting his private business venture for a traditional township with mixed ownership and public spaces. Like the department store owners in major cities, he organized a full schedule of activities: art shows, concerts, dances, holiday fireworks, even pet parades. The plaza was a hit, and the first true suburban mall.

The Great Depression and World War II put the brakes on suburban development, but the end of the war signaled the start of a great migration of white Americans to the suburbs, fueled by car ownership, road building, government mortgage and tax policies, the baby boom, an expanding economy, and innovative products and good marketing by the real estate industry. J. C. Nichols's model of embedding a full-service shopping center at the heart of a new suburban housing tract to help sell homes to buyers who might otherwise worry about isolation from the city became the standard. Commercial developers went one better, realizing that regional shopping centers, deliberately sited at the confluence of regional highways with easy car access and parking, could become magnets for large numbers of customers. In 1945, no more than a few hundred shopping centers of any size had been built, generally similar to Country Club Plaza, but gradually moving away from the romantic historic styles and village-type clusters toward modernist architecture and fewer, larger buildings. In 1950, outside of Seattle, architect John Graham made a leap forward with Northgate Center, two parallel strips of shops anchored by a department store—a downtown stalwart convinced to open an outlying location to reach its retreating customers. The next year, Shopper's World opened outside Framingham, Massachusetts—a similar big strip pedestrian mall, surrounded by parking lots. The effort was hugely costly and risky, as it was difficult to assemble land that was correctly sited, cheap enough but not too far out, and then to attract tenants and arrange financing. Shopper's World failed after several years. But the promise of the concept was too powerful to ignore.

In 1954, Northland Center, designed by architect Victor Gruen, opened in the suburbs of Detroit, with a unique, modernist cluster plan, incorporating 80 acres of landscaping, parking for 10,000 cars, and full climate control with central heating and air-conditioning—the first mall to claim it. Gruen, an Austrian Jew originally named Grünbaum, known at home in Vienna for luxury shop designs on one hand, and writing and acting in socialist plays on the other, had arrived in New York City in 1938 at the age of 35 after fleeing the Nazi annexation of Austria. He found work through friends, including on the New York World's Fair's Futurama exhibit. Soon, he was back to designing stores, and eagerly writing and speaking about his theories of architecture-driven retail environments. After he relocated to Beverly Hills, California, with his second wife and partner, Elsie Krummeck, his practice grew to national prominence. In 1943, the magazine *Architectural Forum*, projecting "the complete redevelopment of the city after the war," published the plans it had commissioned from prominent architects to remake the city of Syracuse, New York, as a model postwar American urban center. The editors provided a "master plan" dividing the city up into three zoned districts of separated functions: residential, commercial, and industrial, with superblock apartment buildings arrayed around superhighways and parks in canonical Corbusian fashion. Mies van der Rohe had been chosen to design the art museum, the Viennese modernist refugee Oscar Stonorov and his protégé Louis Kahn the hotel, the California modernist Charles Eames the city hall, and William Lescaze a gas station envisioned as a multiple-use destination, with a glass-walled restaurant overlooking a freeway. The editors chose Gruen and Krummeck to design a shopping center, meant to serve "the neighborhood." Instead, the partners proposed a single, massive project to serve the entire city of 70,000. It was to be located at the intersection of two highways, its stores and other functions grouped in concentric circles, with the inner building made of glass, inside which "the shopper may proceed from store to store without leaving the actual building"—a giant conceptual step toward the completely enclosed mall. The center would having 28 stores, including a department store, a theater, markets, cafés, shops, and a gas station. It would also offer a series of functions not directly related to the "Stores," but to another dimension that Gruen called the "Communal," including a post office, library, nursery school, pony ride, auditorium, public clubhouse, exhibition hall, and information booth.

Following the success of Northland, Gruen next opened the Southdale Center in suburban Edina, Minnesota, in 1956. In it, he tried out new solutions to two of the biggest problems the mall developer faced. In response to Minnesota's too-hot summer, too-cold winter, Gruen completely enclosed the mall, installing climate-control systems to maintain a year-round temperature of 72 degrees Fahrenheit. To limit the sprawl of the mall itself, now grown so big that walking from end to end would be daunting, he split it into two floors, linked by open escalators in a central courtyard, which was ostentatiously landscaped with tropical plants and fountains and called the Garden Court of Perpetual Spring. He had achieved complete introversion, erasing any awareness of the outside world—as well as, finally, his and Krummeck's Syracuse vision of 13 years earlier.

Gruen was an eager promoter of his innovations, and of the inevitability of enclosed malls: "In providing a year-round climate of 'eternal spring' through the skill of architects and engineers, the shopping center consciously pampers the shopper, who reacts gratefully by arriving from longer distances, visiting the center more frequently, staying longer, and in consequence contributing to higher sales figures." To further sway developers to pay for the additional costs, he produced charts of the time and distance a person would be "willing to walk, under varying environmental circumstances," to prove "empirically" that the investment in air-conditioning would pay off. In his calculations, the enclosed, climate-controlled mall won by a factor of 10 over unembellished city streets.

Gruen's vision wasn't unique—enclosed malls were a national trend, within a national tidal wave of suburban shopping center construction. In 1956, Mondawmin Center, by the planner-turned-developer James Rouse, opened near Baltimore, with handsome modern landscaping by Dan Kiley. Gruen opened a second suburban Detroit mall, Eastland Regional Shopping Center, in 1957. In 1958, there were 2,900 shopping centers in the United States. Thousands more would follow, including scores by Victor Gruen. He was confident that he had hit on a perfect formula, his "recipe for the ideal shopping center":

> Take 100 acres of ideally shaped flat land. Surround same by 500,000 consumers who have no access whatever to any other shopping facilities. Prepare the land and cover the central portion with 1,000,000 square feet of buildings. Fill with first-class merchandisers who will sell superior wares at alluringly low prices. Trim the whole on the outside with 10,000 parking spaces and be sure to make same accessible over first-rate under-used highways from all directions. Finish up by decorating with some potted plants, miscellaneous flower beds, a little sculpture, and serve sizzling hot to the consumer.

There were, of course, variations: the layout might be a "dumbbell" with one, two, or even three major department store "anchors" at either end of a pedestrian strip, or arcade, of smaller shops; or it might be a cluster of separate buildings, or a pinwheel of them. But the entertainment aspect was always crucial: Gruen was at pains to imitate something of the atmosphere of a town square by including benches, light posts, water features, landscaping, and programs of various kinds, just as Nichols had at Country Club Plaza, to encourage people to come, linger, and watch. He boasted that his designs attracted the elderly, children, and their parents alike, simply to spend time, which increased sales per square foot. Did Gruen influence Walt Disney in the design of Disneyland, which opened in 1955? It is very likely, as Gruen's malls were nationally publicized, and Disney was an avid student of advances in urban design. It's equally likely that Disney influenced Gruen, as both were part of the same movement: creating urbanistic spectacles in suburbia, part nostalgia, part futurism, using all the tricks of the Paris arcades, updated with Muzak, subtle lighting, and reflective surfaces, to entice twentieth-century flaneurs. Indeed, observers of the shopping mall coined a psychoanalytic-sounding term for the Gruen formula: "the Gruen Transfer," or the transformation of a destination shopper looking for a particular item into an impulse shopper drawn unpremeditatedly into the mall by its movement and spectacle.

Most important, a successful mall had to cater to motorists, as Gruen was careful to point out: "Be sure to make same accessible over first-rate under-used highways from all directions." Other federal policies only drove the nails deeper into traditional urbanism's coffin, while stacking the deck in favor of suburban housing and retail. Tax increment financing, an arrangement in which future property taxes were dedicated by local governments to cover the costs of immediate public investment in infrastructure to support private development, was one such taxpayer subsidy; regressive sales taxes were another. Federal financial rules capping the amount of commercial space allowed in a building (just 15 to 20 percent) pushed banks away from lending to low-rise, mixed buildings—the kinds that made up traditional main streets.

If the formula, Gruen's or another's, was followed correctly, a regional mall was, in the words of the national developer Edward DeBartolo, said to have at one time owned one-tenth of American mall space, "the best investment known to man." Between 1950 and 1970, fewer than 1 percent of malls failed.

Victor Gruen's malls were in demand: in 1961 alone he opened three, including the Cherry Hill Shopping Center in suburban Camden, New Jersey (with James Rouse as the client); in 1962, three more opened. By 1963, there were 7,100 shopping centers in America, at the vanguard of, and without doubt enabling, white America's move to the suburbs. And yet Gruen increasingly railed against that very landscape. In speeches and books, including *The Heart of Our Cities: The Urban Crisis; Diagnosis and Cure* of 1964, he condemned suburbia itself as a "boring amorphous conglomeration which I term 'anti-city,'" a "chaos of congestion," a "cancerous growth" of "spread, sprawl and scatterization." He analyzed its tendencies toward racial segregation and isolation of people from direct contact with others, and enumerated its typologies with astute, descriptive names such as "transportationscape," "suburbscape," and "subcityscape."

Curiously, he went on to argue that the cure for suburbia's ills was in fact the regional shopping center itself, if it was designed with the sorts of "communal" functions he had proposed for the Syracuse plan: "Shopping centers have taken on the characteristics of urban organisms serving a multitude of human needs and activities," he wrote, comparing them to the "community life that the ancient Greek Agora, the Medieval Market Place and our own Town Squares provided in the past." What he envisioned were multiuse cities in miniature, re-creating a form of urbanism in the suburbs to alleviate the suburbs' rejection of the city. The mall would be "a social, cultural and recreational crystallization point for the up-to-then amorphous, sprawling suburban region . . . a new heart area. Medical office buildings, general office buildings, hotels, theaters, auditoria, meeting rooms, children's play areas, restaurants, exhibit halls have been added to an environment that originally served only retail activities."

Gruen's ideal mall was a direct critique of American society: by reattaching the "Communal" to the "Stores" he would return suburbanites to the public life they had abandoned when they abandoned the city: "In the sound city there must be a balance between the pleasures and comforts of private life and the values which only the public phases of life can offer. We in the United States have upset this balance in favor of private life." He praised Jane Jacobs's *Death and Life* and offered his own systematization to go with her four conditions: his "three qualities or characteristics that make a city" were "1. compactness, 2. intensity of public life, 3. a small-grained pattern in which all types of human activities are mingled in close proximity." His writing brimmed with nostalgia for the traditional forms of European cites—so at odds with the modernist program of his own contribution to American cities. In *The Heart of Our Cities* he even offered a romantic quote from the 1833 Balzac novel *Ferragus: Chief of the Devorants*: "But ah, Paris! He who has not stopped in admiration of your dark passageways, before your glimpses of light, in your blind alleys deep and silent, he who has not heard your murmur between midnight and two in the morning, does not know your true poetry nor your strange and vast antitheses."

In 1956, Gruen had been invited by an electrical utility to reimagine the downtown area of Fort Worth, Texas, and had submitted a plan for just such a redevelopment scheme: the entire downtown of tall towers would be isolated, surrounded by freeways, ramps, and parking garages. Inside, circulation would be pedestrian dominated, with several streets closed off to cars, and others covered with overhead pedestrian bridges and plazas. It was in almost every way the standard Urban Renewal scheme, except for its massive size, focus on integrating functions, and the militant filtering out of cars from the pedestrian-friendly center. This was enough to elicit praise from Jane Jacobs, whose review Gruen quoted at length in *The Heart of Our Cities*: "The plan by Victor Gruen Associates for Fort Worth . . . has been publicized chiefly for its arrangements to provide enormous perimeter parking garages and convert downtown into a pedestrian island, but its main purpose is to enliven the streets with variety and detail. . . . The whole point is to make the streets more surprising, more compact, more variegated, and busier than before—not less so."

The Fort Worth plan wasn't built: it would likely have proven far more expensive than anything actually attempted under Urban Renewal to that point. And yet the availability of federal money through Urban Renewal programs motivated cities, especially smaller ones, to seek funding for "revitalization" projects for their ailing downtowns, unable to compete with the suburbs (and malls like Gruen's). Gruen, ever alert to opportunities, pioneered the genre. In 1957, he was hired by the city of Kalamazoo, Michigan, to rework its downtown. His solution was to combine street closures to make pedestrian-only malls and the rerouting of cars away from the area to minimize traffic with landscaping: patterned and colored paving, fountains, trees, benches, lampposts, kiosks, and so on. The Kalamazoo downtown opened in 1959 to much acclaim. Gruen was then hired by a long list of cities to do studies, and he built pedestrian malls in Honolulu, Hawaii (Fort Street Mall), and downtown Fresno, California. He was well aware of the shortcomings of this "on the cheap" approach of simply closing streets and installing "street furniture" such as benches and lampposts, but the low costs removed barriers to action, and by the late 1970s more than 200 US and Canadian downtowns had some form of pedestrian mall. But with few exceptions, none could compete with larger suburban malls, which lured people with their selection, pricing, perception of safety, and, paradoxically, higher densities of shoppers. In 1954, suburban malls surpassed metro downtowns in retail sales for the first time. In 1970, there were 13,000 shopping centers in the United States. By 1976, the suburban branches of major department stores accounted for 78 percent of their sales.

In 1968, Victor Gruen retired, embittered by his failure to realize his high ideals, to his native Vienna. His firm, Gruen and Associates, eventually built 45 shopping centers in the United States, with 16 others unbuilt, plus 14 department stores outside malls—a total of 44.5 million square feet of shopping. He was routinely referred to in the media as "the father of the mall"—a label he angrily rejected. In a 1978 London speech (later published in article form as "The Sad Story of Shopping Centers"), he accused American developers of perverting his "environmental and humane ideas" by keeping only "those features that had proved profitable," multiplying the number of developments with a simultaneous "tragic downgrading of quality." "I refuse to pay alimony for those bastard developments," he spat. And yet Victor Gruen's patrimony could not be denied.

Nor could his vision of bringing the mall back to the city, even in the face of suburbia's relentless rise and the hollowing out of downtown. Many were inspired by this vision, including James Rouse, who developed the planned town of Columbia, Maryland, and who was Gruen's client at the Cherry Hill mall in New Jersey. Rouse had been captivated by the Fort Worth plan, calling it "the most magic plan and the largest and the boldest and the most complete dealing with the American city I have ever seen. . . . It is a wonderful image of what the city could become." He got his chance to do an urban project when the architect Ben Thompson, a former chairman of the Harvard Graduate School of Design and onetime partner of Walter Gropius, brought him an opportunity in the old downtown of Boston, Massachusetts. It was a six-acre site adjacent to Faneuil Hall, the famous colonial market building, on the edge of the financial district and just east of the new city hall (a startling concrete brutalist homage to Le Corbusier by architects Kallmann, McKinnell and Knowles, finished in 1968). The site contained three derelict warehouses: the 535-foot-long Quincy Market building, a venerable but badly dilapidated nineteenth-century market with a flooded basement, rotted framing, and a corroded copper dome, flanked by two other five-story buildings in similar condition. At first look, it wasn't promising, but Rouse saw potential: thousands of people worked nearby, and thousands more walked through the site following the Freedom Trail of Boston's historical sites. Influenced by Gruen and by the success of Ghirardelli Square, an outdoor shopping center opened in an abandoned chocolate factory on the San Francisco waterfront in 1964, Rouse envisioned a historical-style market hall filled by local vendors selling crafts and food. He saw the tourists already in the area as the largest market segment and planned to provide entertainment as a core offering.

The problem was, no banks were interested in lending to build the project. There had been no new retail development in Boston in 20 years, and the national business consensus was that suburban malls were the only game in town. The mall tycoon Edward DeBartolo, who at one point owned almost a tenth of US mall space, said in 1973: "I wouldn't put a penny downtown. It's bad. Face it, why should people come in? They don't want the hassle, they don't want the danger. . . . No individual or corporate set-up can make a dent on those problems. So what do you do? Exactly what I'm doing, stay out in the country, that is the new downtown." Yet Rouse believed that the pendulum had reached its farthest point, and that the revival of central cities was imminent: "The American dream for millions and millions of young Americans is no longer a quarter-acre lot and a picket fence. It's rehabilitating a house in the central city, or buying one that's rehabilitated, typically [by] two people who are working. . . . This kind of household doesn't find its life best lived out in the suburb, but in the center of the city with its convenience and vitality."

With intense lobbying to bring civic leaders and their persuasive power to his side, Rouse secured financing—only for the first half of the plan, and only by combining contributions from 11 banks. The first phase of Faneuil Hall Marketplace opened in 1976, an indoor-outdoor complex of shops and stalls, enlivened with carts, kiosks, performers, and bright colors. Rouse drew lessons from his favorite developer, Walt Disney, and his Main Street U.S.A., in seamlessly combining retail, entertainment, and tourism: "Shopping is increasingly entertainment and a competitor with other entertainment choices," he wrote. "In a circumstance of delight, it gratifies a need that might otherwise be met with a trip to New York, or a weekend at the beach." And, like Main Street U.S.A., Faneuil Hall was a careful balance between historical restoration and modern commercial techniques—a kind of "retroscaping" somewhere between the "living history museum" of Colonial Williamsburg, Virginia (itself a hybrid between original historical fabric and aggressive retailing), and Disneyland (with its nostalgic appeal frankly combined with crass commercialism). Critics complained that it was a historical fake and that its urbanity was merely staged for dollars. Rouse countered that the "real" city and commerce were in no way exclusive; indeed, that business could rescue the city: "Profit is the thing that hauls dreams into focus."

Faneuil Hall was an immediate success, and a new genre of shopping center was christened: the festival marketplace. Rouse's firm followed up with Baltimore Harborplace in 1980—another well-publicized coup. The developer graced the cover of *Time* magazine on August 24, 1981, under the headline: "Cities Are Fun! Master Planner James Rouse." Next were New York City's South Street Seaport, in 1983–85, Washington, DC's Union Station, in 1988—a revamping of Burnham's cavernous neoclassical space—and the Waterside, in Norfolk, Virginia, in 1983. Each of them helped to revitalize their respective downtowns, but each also struggled in the face of persistent difficulties in reestablishing a middle-class beachhead in the cities, beyond tourism. The Rouse Company moved on to build festival marketplaces in Toledo, Ohio; Flint and Battle Creek, Michigan; and Richmond, Virginia—all of which eventually closed. The company had more success overseas, beginning with Harbourside, in Sydney, Australia, in 1988, and Tempozan Marketplace, in Osaka, Japan, in 1992. In America, downtown's time had not yet fully arrived.

In 1990, there were 36,515 shopping centers in America. This peak of success was also the point of saturation, and the beginning of a period of crisis for the mall. New competition came in the form of outlet malls, often built completely outside cities on discount land alongside new highways, selling steeply discounted merchandise, and "big box" stores like Walmart and Home Depot, either standing alone before a vast parking lot or grouped together in a "power center"—a kind of gargantuan return of the box-behind-parking

1930s one-store strip shopping center. At the same time, home shopping on cable TV and an expanding Internet bringing a proliferation of online shopping options made a trip to the mall just one choice for consumers—and frequently it was the most time-consuming option. Dining out with friends, seeing a movie, taking vacations, going to the gym or health club, or just staying home and watching cable or a movie on any of a steady stream of new devices—VHS, DVD or BluRay player, or DVR—all took a bite. The time shoppers spent in malls declined by half from 1980 to 1990. By 1995, American women were spending 40 percent more on vacations and 21 percent more on dining out than they had a few years before. Mall vacancies rose: in 1995, vacant big-box space in the Chicago region topped 12 million square feet. By the hundreds, malls began to fail in the United States. Back in the Valley, the Sherman Oaks Galleria dwindled down to a few small shops, before closing in 1999, leaving its owners searching for a new formula for success. To fail would mean the enclosed suburban mall would be the endpoint of a centuries-long progression of shopping architecture. Fortunately, a breakthrough to a new paradigm had been realized, just a few hours south of LA on the Interstate 5 freeway.

In 1977, the veteran suburban mall developer Ernest Hahn found himself with an unwanted downtown problem. He had his eye on some juicy sites in the fast-expanding San Diego region, including in Mission Valley, where hundreds of millions of dollars of new freeway construction was set to converge, and in Escondido, a booming northern suburb. But city of San Diego mayor Pete Wilson (a Republican who went on to become governor of California from 1991 to 1999) held the key to one of the real estate options Hahn needed, and Wilson wanted the developer to solve the mayor's own downtown problem first—with a shopping mall. The city's site covered six abandoned blocks of the dilapidated downtown area, a classically "blighted" mixed-use district between the old waterfront to the west; the concrete-and-glass towers of a new, Urban Renewal–bought financial district to the east; and a US Navy base's sprawling docks to the south. On the hill to the north, the tiled cupolas of Bertram Goodhue's dream city at Balboa Park punctuated a skyline of eucalyptus trees. Due in no small part to the presence of the navy, the area was heavy with pawn shops and X-rated video stores. When Hahn, a formidable salesman, brought the chairman of the Montgomery Ward department store group to see the opportunity there, a homeless man urinated on his shoes. Clearly, the standard suburban mall wasn't going to work.

Hahn made a phone call—to Jon Jerde, an architect and longtime executive at the firm of Charles Kober and Associates, with whom he had collaborated on several projects. Jerde had spent the previous 13 years at Kober and its predecessor firm designing conventional California malls, completing his first one in 1968 and continuing notably with the Northridge and Los Cerritos malls in 1971, the Glendale Galleria in 1976, and Hawthorne Plaza in 1977, all versions of the typical dumbbell, enclosed plan. Trained in fine arts and engineering before moving to architecture at the University of Southern California (USC), Jerde worked up to director of design, then executive vice president at Kober, but remained frustrated with his career. "Retail is the bottom of the bucket in terms of what ambitious architects want to work on," he reported to an interviewer. "So I was very depressed." Earlier in 1977, he had quit Kober to assess his next steps. When Hahn told him that something different was needed to make the San Diego site work, and that he was willing to pay Jerde to come up with it, he jumped, forming the Jerde Partnership with Eddie Wang, a colleague at Kober who had been born in China and raised in Taiwan, and had come to the University of Illinois for architecture graduate school. What they set out to do was reinvent the shopping mall with "a deliberate urban script, a conscious creation of human urbanism"—in other words, to bring the mall back to the central city, by reimagining it as an image of urbanity.

Jerde's vision had many ingredients, but ultimately was rooted in his childhood. Born in 1940 in Alton, Illinois, Jerde has described his family as being "oil-field trash," following a father in the nomadic oil business from place to place. When he was 12, he moved with his mother to Long Beach, California, a solid, working-class city on the coast south of Los Angeles, built on oil fields, with a major port, a naval air station, and a McDonnell Douglas aircraft plant dating from World War II. Young Jon was something of a loner, drawn to architecture and city design from an early age, wandering the back alleys collecting bric-a-brac with which to make "fantasy cities" in the garage of his rented house. He spent much of his free time at the Pike, an old-school amusement zone on the seafront packed with roller coasters, rides, game arcades, shooting galleries, bars, cabarets, and tattoo parlors. He was drawn by the color, noise, and cheap entertainment, and by the crowds, finding, he would say years later, "a wonderful warmth and sense of belonging among all these anonymous folks." In the course of his architectural career, it was this "communality," as he calls it, that he despaired of losing in the featureless voids of the modern mall and hoped to recapture in the design for what would become Horton Plaza in San Diego.

During a yearlong travel fellowship he won in 1963 as an undergraduate at USC, he wandered through Europe looking at traditional urbanism and was especially captivated by the hill towns of Italy, by the liveliness of their narrow, twisting streets, filled with people on a variety of errands, whether working, shopping, or passing through, but also conversing, watching, and enjoying. The hill town street, like the Pike, but unlike the postwar American suburban street, was a venue for social interaction as well as commerce. The shape of the place was more important than the buildings, which could be highly varied, yet always growing from a common context, history, and tradition. As an architect, Jerde became obsessed with how the "re-invention of communal experience" that had been lost in the modernist, car-dominated city could be achieved through "experiential placemaking," in his words.

In Southern California, he knew of older examples of a similar, pedestrian-scale "villageness" because of the way the region had originally been settled and developed around many small, dispersed centers rather than one central downtown. And he knew of particular places where Mediterranean architectural traditions had been successfully combined with shopping: Santa Barbara's downtown, for one, a partly historic, partly "retroscaped" shopping and dining district shot through with "paseos," shop- and restaurant-lined pedestrian walkways and arcades that wander between the streets like a "necklace," in Jerde's words. It pointed the way toward a method of "scripting the city . . . to create urban theater, providing the gentle background to what goes on in terms of the encounters as you move through a city series of interconnected projects, a trail of breadcrumbs that lead you through the city." Despite his frustrations with retail architecture, he became convinced that it was shopping that offered the best opportunity to build communality: "The shopping center is a pretty pathetic venue to deal with broad-based communalizing, but it is all we've got and it's the pump-primer in America," he wrote. It was a kind of apology, but also a confident claim: "the consumption addiction is what will bring people out and together."

The architectural solution he and his team came up with for the Horton Plaza site was an "armature," a pedestrian street three levels high, open to the sky, with entrances aligned with the city streets outside, cutting diagonally through the core of the six blocks. The street would be lined by a series of buildings to house all the elements of a regular mall: hotel, theater, shops, restaurants, offices, cinema, and anchor department stores—not two, or three, but four; and a series of spaces designed to encourage movement: a galleria, a terrace, an open courtyard. It all added up to an elongated, multilevel version of the festival marketplace. The armature consisted in plan of two long, shallow arcs, which reversed direction near the middle, channeling people from one side to another the way a river switches its main current from bank to bank as it flows downstream. In keeping with the postmodernist sensibility ascendant in architecture at the time, with its free use of historicist decoration and stagey, populist borrowing from lowbrow commercial design, the team sought out examples of local buildings to integrate. "In Horton Plaza's 40 acres we tried to draw upon all the elements that make up San Diego," Jerde wrote, "from Spanish revival churches all the way up to modern Art Deco buildings, up to very bad contemporary mall buildings and everything in between." He believed that the technique wasn't mere pastiche, but a way to distill the "personality or persona" of a place, the "collective fantasy" its inhabitants have of it—"a primary perceptual method by which people form a bond with their home." Done right, "these qualities are amplified, like the theater amplifies, so they feel bigger than life. You experience the place at a higher volume, as the ideal fantasy of what you believe this place to be."

To the San Diego–derived themes were added elements Jerde had sketched in notebooks during study trips to Italy taken during the design of the project: a triangular building jutting into the pedestrian street at the point where the arcs reverse, an arch across the space, a rectangular plaza, a series of triangular windows on a high wall, a row of arches overhead. These details lend a distinctly Mediterranean, even classical, tone to what is otherwise an architectural riot. The elements are thrown together in layered, staggered angles, with terraces and balconies, escalators, stairways, pedestrian bridges, nooks, and that postmodernist signature, the "rhetorical" colonnade, extending mysteriously into space, holding up nothing. On top of it all is liberally sprinkled the full panoply of Gruen-Rouse retail theater paraphernalia: fountains, art and sculpture, signs, flags, and banners. Color is everywhere: splashed over the stucco surfaces—the triangular building is checkerboarded in white, black, and red, as if covered in marble like the Duomo in Florence; lights project color washes onto walls; fabrics are suspended in all directions. The firm's recent experience designing a unifying theme for the 1984 LA XXIII Olympics helped refine the tricks. Paired with a graphic design firm, the Jerde Partnership had come up with a "kit of parts" to be applied to the 75 Olympic venues spread all over the city: banners, columns, fences, canopies, tents, towers, and gateways made of fabric, cardboard tubes, metal pipes, and scaffolding, all modular, all soaked in brilliant, hot colors, and cheap, to stay within the games' tight budget. It was part and parcel of "a spirit afloat in Southern California about making cheap architecture," according to Jerde, "about making things out of prosaic materials like chicken wire, chainlink, and temporary structures." Frank Gehry's contemporaneous work comes to mind.

Horton Plaza opened in 1985 to accolades, consternation, enormous press attention, and crowds—25 million visitors in its first year. Its success catalyzed the redevelopment of the entire area, renamed the Gas Lamp Quarter, and major investments in a light rail line to Tijuana, just across the border in Mexico, a convention center, and a baseball stadium. For the Jerde Partnership, it was gratifying proof that the recipe of reviving the central city with a top-off, urbanistic retail experience worked. And another shopping genre was born: the "experience mall."

In the wake of Horton Plaza the firm grew quickly, with many of their new jobs involving the reinvention of traditional enclosed malls. For the remodel of Westside Pavilion in LA in 1985 it opened the building to the street using passageways, storefronts, and bold signage, and in 1989 transforming the staid Fashion Island mall in Newport Beach into a pseudo–hill town of intimate passageways and arcades, à la Santa Barbara, all surrounded by seas of parking—reversing the Horton formula in a bid to "rejuvenate the complex using urban values in the heart of suburbia," in Jerde's words. It also got the chance to show that the alchemy could, at sufficient scale, as easily be boxed up and put inside a conventional suburban sprawl mall to turn it into an entertainment destination in its own right. For the Mall of America, in Bloomington, Minnesota, Jerde was given a 96-acre site at the intersection of two big highways in the suburbs of the Twin Cities of Minneapolis–Saint Paul, on which the firm put the world's biggest dumbbell mall, with a quarter-million feet of retail space on four levels, occupied by more than 500 stores and movie theaters, a hotel, an underground aquarium, and a full-blown indoor theme park at the center. Its principal owner, Melvin Simon, gushed during the construction that it was "going to be like going to Disneyland." Jerde later said of his creation, "What they wanted was four malls bolted end to end, so it was a piece of shit. [But] I went in opening day, and I went, wait a minute, this isn't so bad. Not the design of it, it's generic. This isn't a shopping mall anymore. This is generically something else. This is a strange new animal here that, if you learn to do it right, could be off-the-wall, I mean really fucking great." After its opening in 1992, a claimed 40 million visitors each year agreed with him. The Mall of America has consistently attracted more people every year than Disneyworld, Graceland, and Grand Canyon National Park put together. Here Jerde was obeying, and demonstrating, Reilly's Law of Retail Gravitation—which states that, "all other factors being equal, shoppers will patronize the largest shopping center they can get to easily"—and proving in the process that the Jerde formula wasn't tied to a city location, but about creating the sensation of a certain kind of urbanistic and communal environment, fulfilling Victor Gruen's visions for a mall-centered, urbanistic suburbia. "Communal experience is a designable event," Jerde wrote, wherever the site may be.

For the sprawling Universal City property in North Hollywood, the Jerde Partnership created a master plan, completed in 1989, intended to connect the three major existing public elements: LA's second-most-visited theme park, a concert amphitheater, and an 18-screen cinema. The plan combined the entire complex into a circular, domed hill town made up of nine distinct districts, each themed for its dominant activity: Canal Street echoed the adult nightlife of New Orleans; Main Street was a toned-down version of Horton Plaza, Urban Village a New Urbanist–lite office and retail zone complete with a "Spanish steps" feature, and so on.

Universal's owners shelved the master scheme, but asked Jerde to come up with a distillation of its concepts to link the three major elements to the parking garages with a retail development. The solution was Universal CityWalk, opened in 1993: a 1,500-foot-long "street" in two parts converging on a central space domed over with open steel canopy and "activated," as designers like to say, by a "dancing," walk-through fountain devised to enchant children, and therefore their families watching them, so creating a milling, lingering concentration of people. CityWalk takes the architectural mishmash of Horton to a new level. A procession of dissimilar building facades (the retail tenants are encouraged to design and change their own storefronts) jockeys at all angles in and out of the street. They seem hung, like pictures in a long gallery, from the central passage or armature: a welter of surfaces, details, and colors, overlaid with neon signage and what Walt Disney called "weenies"—features meant to draw the eye: a giant cutout of King Kong suspended overhead, a life-size fiberglass dragon, a car crashing through a wall, an electric guitar two stories high. Music, lighting, street performers, and surging crowds of people give the place the excitement and motion of a distinctly surreal cinematic street scene—say, from the movie *Blade Runner*, the classic imagining of a dystopian-utopian future Los Angeles. It is a visual tumult, to be sure, but this doesn't begin to capture the experience—and this is the point, said Jerde, "Our stuff isn't supposed to be visual. It's supposed to be visceral."

Most critics hated CityWalk. Eight years after Horton Plaza, the mood of the architectural mandarins had changed, from indulging interest in postmodernism's populist follies to disdain for its perceived cheapness and lack of rigor. One grumpy contributor to the *Harvard Design School Guide to Shopping*, Daniel Herman, complained of Jerde: "He throws large amounts of architectural matter at the shopper: countless turns and counterturns, unlikely ramps stuck to soffits, thresholds over thresholds that dislodge the visitor of certainty, sending the 'keyed-up'"—a term Jerde had used to describe visitors excited by the architecture—"over the top and into a drone state of consumption." Yet even as he berated Jerde he acknowledged the architect's enormous influence, even granting it a name, the Jerde Transfer: "Whereas the Gruen Transfer produced a new dawn of shopping by introducing an abstract minimal context, its successor paradigm, the Jerde Transfer, has brought shopping to an environmental climax by taking the entire discarded repertoire of architecture and returning it as farce." The last clause makes clear the critic's gravest charge—that Jerde rejected the abstract minimalism required by modernism, which had returned to architectural correctness, for an unserious historicism. The bad reviews reaffirmed the establishment's judgment that "retail architecture is below the trash can," as Jerde had long understood.

Beyond qualms about the pastiche quality of the architecture is the serious charge that Jerde's "artificial cosmos" is nothing more than fake urbanism. But CityWalk was designed not as a model city, but as an entertainment, shopping, and dining business proposition, and it was—and remains in 2016—massively successful, filled with people, locals and tourists alike, bridging the theme park and the theaters, constituting a regional attraction in itself. Jerde's place at the head of the vanguard of entertainment architecture was confirmed, again with perfect timing: the firm was perfectly positioned to help the casino owners of Las Vegas learn from the mall as they sought to remake Sin City into a family-oriented resort destination. In the early 1990s, Vegas's casino operators faced a similar challenge to suburban mall owners, magnified by economic recession and changing demographics as their target audience of young gamblers settled down with children. Jerde was approached by the casino owner Steve Wynn as he was planning a new themed property, Treasure Island, meant to attract families. Eddie Wang recalled that when Wynn discovered that Jon Jerde was a pirate fanatic, Wynn said, "Why don't you come tell me what to do?" Jerde came up with the famous hourly sea battle between a pirate ship and a British frigate in Buccaneer Bay, a man-made lagoon in front of a cheerfully fake and Disney-esque clutter of pirate-themed shopping his team slung at the base of the standard L-shaped slab of the hotel. (The blazing broadsides from the ships recall the noon-hour cannon shot in the Palais-Royal's garden, two centuries earlier.) Next, Jerde was asked by Wynn and other owners of aging casinos in Las Vegas's old low-rise downtown, known as "Glitter Gulch" for its neon signs, to propose a way to compete with the growing number of mega-resorts on the nearby Strip. Jerde's solution was the Fremont Street Experience, built in 1995. The street was closed to cars and roofed over with a canopy 1,400 feet long, 100 feet wide, and 100 feet tall, in the classic shape of a Parisian steel-and-glass arcade, illuminated with a computer-controlled light and sound system on its underside to transform the night sky into a glittering firmament of graphics and virtual fireworks. It has been credited with reinvigorating the downtown area. The firm went on to create a large number of the next generation of iconic Strip entertainment clusters, including the dancing fountains at the Mirage in 1998, the Wynn resort; and the over-the-top Bellagio, with its re-creation of an Italian village set before an 11-acre lake, and including luxury shopping streets, cafés, an art museum, an arcaded conservatory garden à la Paxton, and a retail promenade under a 50-foot domed glass ceiling. Eighty thousand people came to see the spectacle in its first 16 hours.

The transformation of Las Vegas from a gambling mecca to a

shopping and entertainment destination was just the most visible marker of the penetration of the new retailing model into the urban fabric of cities all over the world. Shops built around themed entertainment, in many cases laid out like mini department stores—Hard Rock Cafe, Niketown, ESPN Zone, Chuck E. Cheese, and American Girl Place—anchored a proliferation of third-generation malls built by the Jerde Partnership or alert imitators. Even Disney migrated outside its own theme parks as its spatial business model was adapted, refined, and profitably redeployed by designers like Jerde: the first Disney Store opened in Jerde's old Glendale Galleria in 1987; 10 years later one opened in New York's Times Square, revamped as an entertainment shopping zone. Disney would eventually have 535 stores in 11 countries—a paragon of the globalization of branded space in a new, consumption-dominated world economy. To prove that the line between culture and commerce had been utterly blurred, critics pointed to the virtual takeover of art museums by their stores: by the end of the century, in-house stores accounted for the biggest share of museum income, up to 25 percent, with their average profit margin of 48 percent outpacing regular department stores by 10 percent. The Jerde paradigm, if not the Transfer, has become inescapable.

For his part, Jerde remained unbothered and unrepentant. He saw the vocation of good entertainment-shopping design as not shopping but therapy. "What we've figured out is that place, and its persona, have the ability to trigger aesthetic emotion," he explained. "Our true client is the populace itself. We put people in a popular and collective environment in which they can be most truly and happily alive." To Victor Gruen's messianic urban reform mission he had added the achievement of mental well-being for a good part of the population. "We are like psychoanalysts, uncovering the dreams of our clients and helping to make them come true."

....

During the US recession of the late 1980s and early '90s, Eddie Wang led the Jerde Partnership to Asia, finding clients in Japan, Taiwan, China, Singapore, and Indonesia eager to apply the firm's vision to projects in many cases far larger than any Jerde had worked on in the United States. The firm's breakthrough project was Canal City Hakata, a huge, Horton Plaza–style mall opened in 1996 in Fukuoka, Japan, with a pedestrian street paralleling a watercourse, overhung by a steep, concave cliff of stacked balconies, said by Jerde to have been inspired by the soaring sandstone alcoves of New Mexico's Canyon de Chelly. Immediately successful, it was, like Horton, credited with prompting the revival of a depressed part of the city. At the time the largest private development in Japan, Canal City was soon dwarfed by the firm's other Asian projects. Many of them went beyond shopping malls to include urban design full stop—complete with offices, residences, hotels, transportation, parks, and entertainment venues, all glued together with the kind of outdoor pedestrian shopping environments the firm had pioneered. A template of sorts had been achieved in 1985 with the design of Satellite New Town, a concept for the EuroDisney development outside Paris that was conceived of as a 6,000-acre utopian suburb in the shape of a massive dome, or hill town, mixing retail, entertainment, hotels, spas, and a conference center in a pedestrian-dominated transportation matrix. A visible influence on the plan was Paolo Soleri, the visionary eco-utopian architect in Arizona. Though unbuilt, the plan helped Jerde design complex master plans for a series of enormous, mixed-use microcities, including Rinku Town, Makuhari Town Center, Rokko Island, Roppongi 6–6, the Dentsu corporate headquarters at Shiodome, and the Namba project, all in Japan, and a growing list of others elsewhere, including Dubai Festival City in the Persian Gulf and projects in eastern Europe.

These developments are not malls, but something else—"a strange new animal" in Jerde's phrase—an integration of shopping into the rest of life, of the "Stores" with the "Communal," with people living, working, shopping, and recreating inside a single, planned entity. Are they an example of city-making? One way to know is to ask if they satisfy Jane Jacobs's four conditions for generating "exuberant diversity in a city's streets and districts"—in most cases they do, at least during their open hours.

They have proven that shopping architecture can reinvigorate existing cities as easily as it can eviscerate them by stoking suburban flight. Theorists have acknowledged that "shopping's effectiveness in generating constant activity has made it an indispensable medium through which movement in the city is enacted. Not only has shopping become a basic building block of the city, it has, moreover, become one of the best tools for providing urban connectivity, accessibility, and cohesion." A mall doesn't have to be just a mall.

In 2000, Jon Jerde claimed that "over one billion people a year visit our projects." The influence of those projects has continued to spread, like the proverbial ripples across the pond. After he and his wife sold their shares in the company to a group of the other partners in 2002, the firm has continued on, pushing the total number of built projects under the Jerde name to more than 100 worldwide. Three years later, another set of Jerde partners split off to form 5+ Design, based in Hollywood, now with an enormous and growing portfolio of mega-projects that includes integrated, private minicities, shopping developments, even a new generation of cruise ships. They and many other architects and entrepreneurs, particularly in China and elsewhere in the developing world, are busy designing and building projects that will serve millions, perhaps tens of millions, of people.

But is it enough? We can only hope that shopping design's evolution toward more inclusion and integration continues. Regardless, as long as it is profitable, it will continue to be a major contributor to the environments we inhabit, as it has been for centuries, if not more. Time will tell. In an essay on the firm's influence, the LA architect and critic Craig Hodgetts asked whether "Jerde's artificial cosmos may, in time, attain the dignity of the truly cosmopolitan . . . with the scars and patina of age." Yet age and familiarity are not what make a place truly urban, but its integration into the fabric of the city around it. The question is then, will Jerde's places become, as some previous forms of shopping architecture have, public places as much as private ones—places integral to urban vitality?

6. Malls

Field Guide: Shopping Worlds

Diagnostics

- A collection of retail shops, restaurants, and other vendors built around a pedestrian-street-style environment. Can aim for verisimilitude, mimicking a “real” street, or be unabashed theatrical themed environment.

Examples

USA

- Horton Plaza, San Diego.
- Mall of America, Bloomington, Minnesota.
- Universal CityWalk, Los Angeles.

Japan

- Canal City Hakata, Fukuoka.

Variants

Casino Resort Hotels

- Las Vegas: Treasure Island, the Bellagio, the Palms.

Airports and Rail Stations

- London: Heathrow Airport; King’s Cross and Saint Pancras railway stations.
- Hong Kong: Chek Lap Kok International Airport.

Images:

Horton Plaza, San Diego, California (1985). Architect: Jerde Partnership International
Coolcaesar at en.wikipedia

Photo by Phil Konstantin

Mall of America, Bloomington, Minnesota (1992). Architect: Jerde Partnership International.

Canal City Hakata, Fukuoka, Japan (1996). Architect: Jerde Partnership International.

Shops at the Bellagio resort, Las Vegas, Nevada (1998). Architect: Jerde Partnership International.

Treasure Island Hotel and Casino, Las Vegas, Nevada (1993). Architect: Jerde Partnership International.

Cloud Nine Mall, Zhongshan Park, Shanghai, China (2006). Architect: ARQ.

Mall of the Emirates, Dubai, United Arab Emirates (2005). Architect: F+A Architects.

Chapter [7]

Habitats

Kenzo Tange, Norman Foster, and the Techno-Ecological City

What is a movement? A form of conspiracy? A shoal of fish changing direction in a single flash? A form of trapeze act? An unstable human pyramid? Or simply a crisis that erupts between geniuses to make it unthinkable to go on in the old way?

—Rem Koolhaas and Hans Ulrich Obrist, *Project Japan: Metabolism Talks*

In 1960, Buckminster Fuller and Shoji Sadao proposed "Dome over Manhattan," a two-mile-wide, transparent geodesic dome covering Midtown from the East River to the Hudson, and from Twenty-First Street to Sixty-Fourth Street. It would eliminate bad weather, and, by being kept at a steady temperature, eliminate the cost of heating and cooling separate buildings. It would also reduce air pollution, though it was unclear how. The self-taught engineer and inventor "Bucky" Fuller was by then moderately famous for his futuristic-looking geodesic domes, made of metal bars forming triangles with linked tips in shapes geometricians call icosahedrons. Covered with a lightweight skin, the domes enclose the maximum volume for the least surface area of any structure, and are light, strong, and easy to assemble and disassemble. Though the geodesic dome was invented in Germany in the 1920s, Fuller worked out its math in the late 1940s and patented it in the United States in 1954, the same year he began working with Sadao, a young Japanese architect studying at Cornell University. Since then, thousands had been built worldwide, including many for the US Marine Corps, which was looking for a helicopter-transportable field structure for its soldiers.

The partners were unsure which materials could actually support the humongous New York dome but were nevertheless confident that its cost would be offset by savings to the city. Fuller wrote, "The cost of snow removal in New York City would pay for the dome in 10 years."

It was an astounding idea—but not because it assumed that an unproven technology could save New York City. The year 1960 was after all the apogee of technological optimism, with rising material prosperity and confidence in man's ability to control his environment. Sputnik was launched in 1957; President John F. Kennedy would announce America's mission to the moon a year later, in 1961; and supersonic passenger flight would soon become a reality with the building of the British-French Concorde plane. What made the big dome different was that it took no position on the existing city: neither that it was blighted and must be cleared and replaced with more efficient towers and highways, nor that it should be defended as it was—the opposing poles of the existential urban struggle even then playing out in streets and meeting rooms over the Lower Manhattan Expressway project. Instead, it redefined the city by the extremes of its physical environment—its notoriously bad winter and summer weather—and proposed to control these through advanced, space-age design and engineering. The proposal, which was presented in the form of an image of Manhattan with a gleaming gauze hemisphere placed over its midsection, captured the optimism of the technology-driven utopianism of the era. It lifted the gaze from the street quite literally to the sky, as did another proposal Fuller and Sadao unveiled the same year: "Project for Floating Cloud Structures (Cloud Nine)," in which lightweight geodesic spheres carrying thousands of passengers, heated by the sun's energy, drifted serenely among the clouds. But "Dome over Manhattan" also betrayed a tinge of fear: of threats from above, whether national self-doubt during the Space Race with the Soviet Union or Cold War fear of fallout from the nuclear holocaust that had hung over New York as America's first city for a tense decade. If Urban Renewal was a case of the architectural and engineering professions imagining completely other physical forms for the city, Fuller and Sadao's was another, sharing the program's blind faith in the power of technology to solve complex social problems. The massive dome promised at once the concreteness of intricate geometry and structural load calculations, and a gossamer vagueness. Was it a structure or a metaphor? It couldn't escape being both.

Fuller's collaboration with Sadao was the second US-Japan collaboration in his life; he had met Isamu Noguchi, the American-born, half-Japanese sculptor, in 1929 in Greenwich Village, where Fuller lived. Back from his apprenticeship with the sculptor Constantin Brancusi in Paris, Noguchi was introduced to the older man at Romany Marie's Café on Waverly Place, the famous bohemian watering hole where Fuller gave informal lectures and had decorated the walls in silver aluminum paint. Noguchi sculpted a bust of Fuller, and the two worked on several projects together, including modeling the inventor's Dymaxion Car, over the course of a lifelong friendship.

In 1959, another US-Japan collaboration on the potential shape of the city was in progress not far away at the Massachusetts Institute of Technology in Cambridge, across the Charles River from Boston, where the Japanese architect Kenzo Tange had come to teach for a semester. During that time, Tange traveled to several other US cities including New York, Chicago, and San Francisco, where he was fascinated by the huge urban infrastructure projects being erected, especially elevated highways through the middle of crowded cities. Free from the day-to-day of running his Japanese practice, he had time to think about growth and change and how to merge communications spaces with architecture—by which he meant transportation corridors, such as railroads and highways, and, as it turned out, seaports. He chose to assign his fifth-year seminar students the task of housing 25,000 people in Boston Harbor. The result was a pair of long, linear A-frame buildings curving from the shore out over the water to a series of man-made islands, with housing terraces stacked on each tilted side of the A-frame modules and a highway running through their central void.

The Boston plan seemed consonant with the Urban Renewal taking place around him—the old harborside neighborhoods like the West End being cleared and replaced with expressways and slab housing blocks—and with the dominant influence of Le Corbusier in modern architecture since the 1930s, including in prewar Japan. It was like an aquatic version of the Swiss visionary's Radiant City. But beneath the superficial resemblance, Tange's plan wasn't really about Boston, but about his home, Tokyo. And, on closer inspection, it represented a leap away from Corbusian verities into a mode of thinking, however improbable, that would gradually come to redefine global architectural practice as fundamentally concerned with its technological ability to engage with, if not control, the natural environment.

Kenzo Tange had spent his childhood in Shanghai, China, living on the grounds of the former British settlement there. Returned to Japan, he first encountered the figure of Le Corbusier at Hiroshima

High School in 1930, when a glimpse of the Swiss architect's design proposal for the Palace of the Soviets in a magazine redirected him from his studies in literature and science toward architecture. "Into architecture," he would later recall, "I felt I would be able to pour all my dreams, sensitivity, and passionate enthusiasm." (Of course, by 1931, when Le Corbusier designed his entry with its theatrical, neoconstructivist arch and curved forms, Stalin had already decreed a baroque mélange of neoclassicism as the USSR's official style.) Tange continued in architecture studies at Tokyo's Imperial University in 1935, and upon graduating in 1938, went to work in the office of Kunio Maekawa, one of Japan's leading practitioners. Much of the work was abroad: since the mid-1930s, Japan had been expanding its so-called Greater East Asia Co-Prosperity Sphere by invading China, Mongolia, Manchuria, Thailand, Vietnam, Laos, Burma, the Philippines, and Indonesia, and had set its architects and engineers to the task of planning vast cities in its new dominions, more vast by orders of magnitude than any landscape they had seen at home. From 1938 on, Tange worked on projects in China and elsewhere, and his design for a Japan-Thai Culture Center in Bangkok won an award. It was noted that his designs tended to mix Japanese and Western influences, traditional and more contemporary, in equal measure—though the mixing wasn't encouraged by the imperial authorities. By 1942, the war's course in Asia curtailed building, so he returned to school in Tokyo, where he was drawn to the study of city planning, especially Western classical precedents like agoras and markets. Even as the country was being bombed, the government asked him to consider how to replan Japan's cities in the aftermath.

Tange was 31 when Japan surrendered. He and other architects and planners of his generation who had begun their careers contemplating seemingly unlimited urban canvases in Asia returned at war's end to face their own cities reduced to ash. The bombing had destroyed half of Tokyo, 60 to 88 percent of 17 other cities, and 99 percent of Toyama. On August 6, 1945, Hiroshima was blasted to smoking rubble, and Nagasaki three days later. Tange was placed in the government reconstruction agency and put in charge of the Hiroshima survey team. He drew up a plan to rebuild the flattened city, influenced by Le Corbusier's idea of separated functions, but the urgency of the remaining population's immediate needs precluded a radical rethinking of preexisting patterns. In 1949, Tange's reconstruction plan won approval in the Japanese Diet and progress began on the Peace Park and Plaza, and the Atomic Memorial Museum, a long, concrete slab laid on its side and raised on pilotis. Working under Tange, Isamu Noguchi designed sculptural concrete railings and details for the Peace Bridge.

The Hiroshima work was widely publicized, and Tange was invited to accompany his former mentor Maekawa to the eighth CIAM conference in 1951, in Hoddesdon, England. There Tange met his hero Le Corbusier, Walter Gropius, and other modernist luminaries. Afterward, he toured Europe for two months, visiting Rome to see Michelangelo's Saint Peter's Basilica and other classical buildings, and Marseille to see Le Corbusier's Unité d'Habitation, then under construction. Back in Japan, Tange's career thrived, as he completed his Hiroshima Peace Center (1956), the Kagawa prefectural office (1955–58), as well as several industrial buildings and town halls, crowned by Tokyo's City Hall (1952–57). All of them revealed the influence of Le Corbusier, whose presence must have been nearly palpable while the National Museum of Western Art designed by the Swiss was being built in Ueno, Tokyo, during 1957–59. A more unique building was Tange's multipurpose auditorium in Matsuyama, on the southern island of Shikoku, a shallow, tilted concrete dome 55 meters in diameter, enclosing a single open room, engineered to be earthquake proof. At a time when few of his countrymen were allowed to travel abroad by the American authorities, Tange toured China, Egypt, and India in 1956, where he saw Le Corbusier's Ahmedabad Museum, a brutalist concrete stack raised on pilotis, and Brazil in 1957, as Lúcio Costa and Oscar Niemeyer planned the audacious new capital, Brasília, which would become the largest Corbusian urban-planning scheme ever realized. Here were tangible demonstrations of a new postwar world rising, a moment for emerging countries outside Europe and North America to stake out their own identities and claims—in great part through the cities they were building. For Tange, it must have been intoxicating and sobering in equal measure.

In postwar Japan, the challenge remained how to house the millions left homeless and the millions moving to cities for the first time, as Japan integrated into world markets, its labor-intensive agriculture modernized, and its rural traditions weakened. For the first 15 years it was dependent on the US occupation, with little money aside from the public sector. Not until 1950 with the start of the Korean War and the US military buildup did private building gain steam. The new war, peaking in 1953, turned Japan from a defeated aggressor into an American ally and fueled an enormous economic boom—in turn stoking debate on Japan's role in the postwar world and what role its traditional culture should play in modernity. Reconstruction of its cities was inseparable from the reconstruction of its national identity. A utopian outlook was unavoidable, as the country was building anew, on a virtual tabula rasa, for a new world, with rising technological confidence and new techniques. The sense of limitless possibility was leavened, however, by an underlying warning tone that was, if not dystopian, then catastrophist: beyond the threat of more nuclear conflict in the Cold War standoff in which Japan now occupied a front line, the island nation's perennial natural hazards of earthquakes, tsunamis, and typhoons rendered city building itself precarious, even dubious. Added to this was anxiety about exhausting the supply of buildable land, already scarce in a mountainous archipelago, as the need to conserve farmland from explosive and largely unplanned urban development drove land prices sky-high. By the end of the 1950s, Japanese cities and Tokyo in particular were gripped by a three-pronged crisis of population pressure, disappearing land, and traffic congestion. The rushing force of urbanization continued to be breathtaking: Tokyo grew from 3.5 million people in 1945 to 9.5 million in 1960; the total population of the metropolitan region would rise from 13.28 million in 1955 to 18.86 million in 1964, making it the world's biggest urban area. In less than a single generation, urban Japan had reached its limits and threatened the spaces that sustained it: its farms, gardens, and fisheries.

To the extent that there was a plan, the model followed in rebuilding Tokyo had been one of Western concentric rings, such as the plan proposed for Tokyo in 1956, based on Abercrombie and Forshaw's plan for Greater London, which assumed a central core, surrounded by radial-concentric growth. The problem was that Tokyo was already an incredibly dense ring around a central void—Tokyo Bay, which extended for 922 square kilometers versus Tokyo's 622 square kilometers—with little room to expand outward into the mountains or the sea. But in Tange's office and in his circle there was new thinking, and he revealed it to the world at the tenth CIAM conference in 1959 in Otterlo, the Netherlands. By then the acknowledged leader of Japanese architectural modernists, Tange presented some of his own buildings, but also two much more radical projects by his younger colleague Kiyonori Kikutake. The first was Kikutake's own house, Sky House, which was based on the workings of the biological cell. It consisted of a single, open square room, floating above the ground on piers, with an integrated system for possible expansions: modules called "move-nets" could be plugged in beneath the floor for bathrooms, storage, and children's rooms, with their support services routed through existing exterior compartments or ducts. It was at once futuristic and traditional, with the adaptable open plan and pitched roof of a traditional Japanese house. The second was Kikutake's "Ideas for the Reorganization of Tokyo City": a 1958 plan for a forest of 300-meter-tall, cylindrical concrete towers, each a central core, what its author called a "vertical ground," with living units plugged in—an arrangement again inspired by cells. In its first version, the shore-based towers were flanked by industrial islands on the sea. In its second variant, the whole thing had moved onto Tokyo Bay: 1958's Marine City, a ring with submerged towers for stability, which was movable, autonomous, and climate-controlled. In a series of 1959 drawings Kikutake studied sea creatures as inspiration for a sea-based architecture: jellyfish, water lilies, and sea plants, as well as buoys, spheres, floating cylinders, pylons, and hexagonal columns. They would provide stability by dampening waves and offer space for food production, he explained: "Like sea plants, an expandable chain of alternating balls and cylinders, both made of concrete, harvest food from the sea just by floating."

Kikutake's bizarre sea-world sought to integrate architecture and urban structure with nature's dynamic processes. It was the opposite of Le Corbusier's functionalist zoning, no longer concerned with rectifying the messiness of the old nineteenth-century city, but with coping with a volatile, possibly dangerous environment of the twentieth century by adapting to natural flux, even by imitating nature in concrete, turning the city in effect into one big organism.

In his introduction of the Tokyo Bay plan at CIAM X, Tange explained: "Tokyo is expanding but there is no more land so we shall have to expand into the sea. . . . The structural element is thought of as a tree—a permanent element, with the dwelling units as leaves—temporary elements which fall down and are renewed according to the needs of the moment. The buildings can grow within this structure and die and grow again—but the structure remains." This thinking was in line with concerns that had been raised at the previous meeting, CIAM IX in Dubrovnik, Yugoslavia, in 1956 (which Tange didn't attend), by a faction of younger architects objecting to the vagueness of CIAM's governing Athens Charter. Styling itself Team X, the group, which included England's Alison and Peter Smithson and Holland's Aldo van Eyck and Jacob Bakema, had introduced "mobility," "cluster," "growth/change," and "urbanism and habitat" as discussion topics—without gaining much consensus. In 1959, the Japanese intervention at Otterlo was far ahead of the answers any of the rest of the modern movement could give.

Tange left the conference in September for his semester at MIT as visiting professor (a job probably set up by Gropius, the department chair). It was an inopportune time to be away from home, as preparations were feverishly under way for Japan's first major international conclave, the 1960 World Design Conference, to be held in Tokyo. As an elder statesman in his profession, with the added authority of the illustrious University of Tokyo where he led an elite design studio, Tange was a member of the steering committee, along with other senior figures in a range of fields, including the architectural critic Ryuichi Hamaguchi, the painter Taro Okamoto, and the architects Kiyosi Seike, Junzo Sakakura, and Takashi Asada. In his absence, Tange had left instructions to Asada, the general secretary, to gather a group of younger colleagues and students into a subgroup to present new ideas combining modernism and Japanese elements, and in the process to put Japan on the world's architectural map. Another participant, Hiroshi Hara, then studying under Tange, recalled that Tange "stepped in and said he was going to create a new Japanese modernism by assimilating tradition." Under Kenzo Tange's quiet leadership, Japan's urban future was to be made from its past.

Asada and the architectural critic Noboru Kawazoe recruited, in addition to Kikutake, Kisho Kurokawa and Arata Isozaki, Kange's former students then working in his office; the industrial designer Kenji Ekuan; the graphic designer Kiyoshi Awazu; and the architects Masato Otaka and Fumihiko Maki, among others. Maki, after graduating from the University of Tokyo, had taken master's degrees in the United States from Cranbrook Academy and Harvard, and had taught at Washington University in Saint Louis, before returning to Japan in 1960. The group, "a kaleidoscopic inventory of the Japanese psyche," in the words of Rem Koolhaas, was diverse in talents, training, and temperament. They were young—Kurokawa a mere 26—and confident, dressed in skinny suits and ties, with cigarettes hanging from their lips, meeting at Asada's Tokyo boardinghouse or in coffee shops to discuss Marxist theories, scientific parallels, and their own ideas. Like Kikutake, Kurokawa was already in 1959 working on radical urban alternatives. He had drawn Wall City, a sinuous wall snaking across the landscape, which contained living space on one side and working space on the other, to eliminate time spent commuting. And, in response to the typhoon damage inflicted on Ise Bay that year, he had designed Agricultural City, a massive grid structure raised above farm fields on pilings, through which delicate "mushroom houses" could sprout, leaving the ground clear for crops and integrating the country and the city without sacrificing one for the other.

A set of principles began to emerge in the meetings: the city would extend into new terrains, including the sea, the sky, and artificial ground; there would be continuous renewal, allowing structures to adapt to modernization and natural change (or calamity); and there would be "group form," or the inclusion of spontaneous building and disassembling by the community, outside the control of the designers. Kurokawa wasn't alone in expressing a certain mysticism: "Architecture, which hitherto was inseparable from the earth, is separating itself from it by expanding towards the universe." Many of the projects were inspired by bacteria, with their endlessly repeated, expanding cell structures, and used biological language like "metamorphosis" to describe urban function and structure designed to change over time. The Japanese word *shinchintaisha*, meaning replacement of the old with the new, was adopted—consonant with the Shinto doctrine of death and renewal through a vitalist life force, and with a tradition of impermanence in certain old Japanese buildings, most famously the Ise Shrine, which had been rebuilt every 20 years since 690 CE. Tange and Kawazoe had been invited to witness the reconstruction in 1953 and would publish a book in 1961 entitled *Ise: Prototype of Japanese Architecture*, celebrating the shrine's simplicity, modularity, and use of prefabrication. Similar were the periodic replacement of capsules in Kikutake's Tower City and the function of tides in his Marine City. Kurokawa caught the essence of how far beyond CIAM's rigid categories the new Japanese thinking had moved: the Japanese architects sought "to understand the shift from a mechanical to a biodynamic age." This was the key shift—from a nineteenth-century assessment of the industrial city and its problems to a twentieth-century one of the complex, fluid city, always unstable, always adapting. The English translation of *shinchintaisha* is "metabolism."

Attending the World Design Conference held May 11–16, 1960, in Tokyo, were 227 architects, commercial artists, and designers, 84 of them from outside Japan, including among the architects the United Kingdom's Smithsons, the Netherlands' Jacob Bakema, France's Jean Prouvé, India's B. V. Doshi, and America's Ralph Erskine, Louis Kahn, Paul Rudolph, Raphael Soriano, and Minoru Yamasaki. Tange's group published a book for the event, an illustrated manifesto: *Metabolism 1960: Proposals for a New Urbanism*. In it were contributions from each member, and four essays. The longest, taking up 35 of the 87 pages, was Kikutake's "Ocean City," composed of his Tower City and Marine City projects, plus an amalgamation of them called Ocean City. Kurokawa contributed "Space City," which included his Wall City and Agricultural City with its mushroom houses. Otaka and Maki cowrote "Toward Group Form," which abandoned the modernist quest for "total architecture" and prescribed instead a clustered structural framework inside which people could create their own dwellings, without hierarchy or mastery, allowing "an intuitive, visual expression of the energy and sweat of millions of people in our cities, of the breath of life and the poetry of living." The critic and former editor Noboru Kawazoe contributed "Material and Man," a mystical paean in which he foresaw recovery from the atomic disaster leading to a new era, building from "the unity of man and nature and the evolution of human society into a peaceful state of unity, like a single living organism. . . . Our constructive age . . . will be the age of high metabolism. Order is born from chaos, and chaos from order. Extinction is the same as creation. . . . We hope to create something which, even in destruction will cause subsequent new creation. This something must be found in the form of the cities we are going to make—cities constantly undergoing the process of metabolism."

In Japan, the conference was viewed as an enormous triumph, and a media flexing wings like a new butterfly as the 1960s dawned fed on the big, outrageous ideas presented there and on the confident personalities it discovered behind them. An ecosystem of architectural magazines arose to propagate the phenomenon. Tange himself went on national TV on New Year's Day 1961 to present his Plan for Tokyo—derived from his Boston project and dated 1960, but delayed by work on the conference. His carefully detailed large model showed a new city straddling the bay, raised on piers, with a massive chain of paired three-lane highways, plus a subway, spanning directly across the water, connecting Tokyo prefecture with its neighbors Chiba and Kanagawa. Five million people would live at sea, and another 2.5 million would work in the central axis containing government buildings, hotels, recreation facilities, offices, and transport stations. The model looked like a tree, with its trunk the straight highway spine, its branches the more irregular laterals, and its leaves the building clusters scattered picturesquely along them.

After Tange's performance, other members of the group jockeyed for media coverage, with competing audacious proposals for Tokyo Bay by Kikutake and Kurokawa. Kurokawa joined the provocative nationalist writer Yukio Mishima as a new national media darling, boasting both Western decadence with tailored British suits and cigarettes and resurgent Japanese masculinity with kimonos and samurai swords. None of their megastructural fantasies was taken up, but commissions did come their way, including some steered to them by a colleague now working in the government—all of them fairly modest and integrated into existing urban areas. Tange completed the Totsuka Country Clubhouse (1960–61), the Nichinan Cultural Center (1960–62), and the Tokyo Olympic stadium complex

(1961–64). Kikutake the radical visionary built the eminently practical Izumo Shrine Administrative Building and Treasury (1963) and the Tokoen Hotel in the summer resort of Tottori (1964).

In their first, formative, and most productive years, the Metabolists, as Tange's group became known, developed a set of typologies, which might be listed as floating cities, and partially submerged ones; artificial ground or man-made land supporting new settlements and infrastructure; vertical layering, including extreme examples of "megaforests" of columns and beams holding up aerial cities (the best example being Arata Isozaki's 1960 City in the Air, a multilevel, three-dimensional circulation system built of a combination of Kikutake's circular shafts connected by massive square beams, all filled out with capsules); natural growth systems, allowing for continuous change and regeneration, such as capsules plugged into space frames or columns; and group form, with its allowance of spontaneity and individuality. Key to all these typologies were massive, centralized infrastructure and transportation networks. All shared an assumption of adaptation to their environment through mechanisms of change, regeneration, and self-regulation, most often through "plugging in" prefabricated capsules or modules to provide the flexible, renewable "leaves" on the tree. The ensemble was an effort to engineer cultural resilience in the face of a raft of looming threats, and to bolster national identity at the same time.

There were several paradoxes floating free among the fantastical megastructures. One, that cities could achieve reintegration with nature not by breaking down their massive, artificial formations into smaller portions commensurate with natural systems, the way most reformist ideas had attempted before with greenbelted neighborhoods and garden cities, but by ratcheting up their artificiality and scale and thrusting away from the earth rather than connecting more closely with it. In seeking more organic qualities, the Metabolists turned to more tangible structure. While this may seem counterintuitive in hindsight, in late 1950s and early '60s Japan it may have seemed inevitable, even natural. It may have seemed logical that there was nowhere else to go but bigger if architects wanted to remain relevant as their culture's navigators.

The second paradox was that spontaneous, natural accretion could be induced to settle on a gigantic, determinative structure, forming a living, resilient fabric. The image of the Italian hill town hovered close behind this notion, with its gradual accretion and ineffably perfect balance between small differences and overall consistency. CIAM Team X had attempted to capture this balance with its distinction between "form" (the big picture) and "design" (the unique details). At the scale of the Metabolist visions it was at least a dream if not an oxymoron. Interestingly, the word "megastructure" appeared for the first time in writing in Fumihiko Maki's 1964 essay "Investigations in Collective Form," a proposal for conjuring smallness with big urban design—and thus a pure opposition. Yet this is exactly what Metabolism sought to do: reconcile bigness with smallness, the city with nature, rigid materials like concrete with dynamic change.

Japanese Metabolism got as much press in the West as it did at home. It had clear precedents there, beginning with buildings from Le Corbusier's "organic" or "brutalist" turn in the 1950s: the Ronchamp Chapel (1950–54), the Assembly Building at Chandigarh,

India (1955), and La Tourette Monastery (1957–60). The New Brutalism led from those some distance toward the Japanese avatars, including the Smithsons' work in the United Kingdom, such as Leeds University (1953), Louis Kahn's Richards Medical Research Laboratories at the University of Pennsylvania (1957–60), which Kahn himself presented at the Tokyo conference, and Bertrand Goldberg's Marina City in Chicago (1959–64). In France, Yona Friedman proposed Ville spatiale in 1958, an enormous canopy to stretch over cities. The mood being set, Metabolism's challenge to orthodox modernism was quickly taken up, most memorably by the British conceptual collective Archigram, which published its first pamphlet of colorful, whimsical illustrations in 1961, depicting cities as places of fun and change, built with lightweight structural elements strongly derivative of the Japanese ideas. In 1963 the Institute of Contemporary Arts in London mounted Archigram's *Living Cities* exhibition, and in 1964 the group put out its pamphlet "Plug-in City," a colorful, whimsical "kit of parts" complete with interchangeable megastructural towers, diagonal frames, living capsules, and cranes, drawn in a comic book style. It was followed by "Megastructure Model Kit"—an actual paper construction set of communication ducts, icosahedrons, living-pods, and platforms, to make open-ended, customizable toy model cities. The ironies in Archigram's version of Metabolism ran deep. More "fun" structures were imagined in the European and North American architectural press, like Cedric Price's Fun Palace (1960–65), and the futuristic megastructure increasingly seeped into the popular imagination, providing the model for the elevated sinful city of Sogo in Roger Vadim's sexploitation film *Barbarella* (1968). But Metabolism's influence was also taken in deadly earnest, especially in Britain, where it dovetailed with the New Brutalism that had become almost the required style of institutional building favored by universities and left-wing town councils for social housing.

Buckminster Fuller and Shoji Sadao published "Triton City" in 1967, a scheme originally developed "for a Japanese patron" in Fuller's accounting, as another model of a floating city for Tokyo Bay between the years 1963 and 1966. He wrote: "Three-quarters of our planet Earth is covered in water, most of which may float organic cities. Floating cities pay no rent to landlords. They are situated on the water, which they desalinate and recirculate in many useful and nonpolluting ways." The concept gained some traction, studied by the US government's Department of Housing and Urban Development, the US Navy, and the city of Baltimore for possible deployment for low-cost housing. When Democratic president Lyndon Johnson left office, however, Fuller reported that the government's interest waned. Nevertheless, in his various inventions, including the Cloud Nine spheres, Fuller saw a new, environmentally sustainable urban future close at hand: "While the building of such floating clouds is some years in the future, we may foresee that, with the floating tetrahedronal cities; air-deliverable skyscrapers; submarine islands; sub-dry-surface dwellings; domed-over cities; flyable dwelling machines; and rentable, autonomous-living, black boxes, man may be able to converge and deploy around Earth without its depletion."

Nineteen sixty-seven was the breakout year for more modest, but successfully built examples of the genre. Kenzo Tange's Shizuoka Press and Broadcasting Center, in Shimbashi, was finished, and the Yamanashi Communications Center, in the city of Kofu was under construction. In Scotland, phase one of Geoffrey Copcutt's Cumbernauld New Town Centre, in the works since 1962, was completed—a multifunction megastructure with shopping, hotel, parking, and different kinds of housing, topped by penthouses, and "demountable enclosures" in the architect's words, for expansion—looking like a cross between a futuristic airport and a penitentiary. But by far the biggest flush was the 1967 Universal Exposition in Montreal, occupying an unorthodox site in the Saint Lawrence River consisting of a natural island, an artificial island, and an enlarged sandspit—approximating the man-made landform the Metabolists had dreamed of—all linked by rail and underground metro tunnels. Downtown Montreal, too, had an extensive underground passage system, comprising eight kilometers of shopping malls, pedestrian tunnels, parking garages, and metro stations—a subterranean city protected from the elements with towers sprouting into the air above it. Huge concrete grain elevators towered over the Expo 67 site from the far riverbank, a fact many commentators noticed, and cited as another reason why Montreal was already closer to being a realized Metabolist megacity than anywhere else. The exposition grounds were akin to a near life-size Megastructure Model Kit erected by architects from all over the globe, using most of the attention-grabbing structural parts in the Metabolist toolbox, familiar to engineers but never before used so lavishly in public buildings: the theme pavilion "Man the Producer" was a composition of tetrahedrons, and the pavilion "Man the Explorer" was a complex steel concoction of tetrahedrons, trusses, space frames, and geometric lattices. Intentionally left unpainted so that it would rust, it looked like a ready-made postindustrial ruin, Reyner Banham quipped that "to call it a 'collapsed and rusting Eiffel Tower' was to pay it a compliment." Other pavilions took the form of pyramids (one actually inverted), eccentric tetrahedrons, and tents, fashioned of concrete, steel, aluminum tubing, and coated fabrics. None resembled, even remotely, a conventional piece of architecture. The US pavilion completed the kit of parts: a 5/8 geodesic dome, 200 feet high and 250 feet in diameter, by Fuller and Sadao, its acrylic-covered steel frame enclosing a seven-level structure serviced by a 150-foot-long escalator, the longest ever built, and a mini-monorail line. A catalog of futuristic transport options beckoned visitors: escalators, monorails, trams, hovercrafts, and helicopters, in addition to the subways. The grounds were enlivened by bright colors everywhere, new electronic inventions, rides, art galleries, and musical performances—including by the Supremes and Petula Clark, guests on *The Ed Sullivan Show* broadcast from the grounds. Eyebrows were continuously raised by the hostesses' miniskirts in the British pavilion, designed by the King's Road sensation Mary Quant. The tableau of up-to-the minute youthful fun and futuristic technology and design could have popped right out of the pages of an Archigram pamphlet and perfectly captured the optimism still prevalent in the broader culture in 1967. Expo 67 was the most successful world's fair of the century up to that point, drawing 50 million people; the total population of Canada at the time was 20 million.

In a fair dominated by exciting buildings, the clear star of the show was Habitat 67, a cluster of modular housing erected on the sandspit by the young architect Moshe Safdie, originally from Israel, a project that began as his sixth-year thesis at McGill University in Montreal in 1960, entitled "A Three-Dimensional Modular Building System." When Safdie finished his undergraduate architecture degree at McGill in 1959, he set out with other students on an eight-week study trip around North America and was powerfully impressed by two opposite things: one, "the force of suburbanization—the desire for dispersal outward from city after city, and the ubiquitous dream of individually owning one's house and garden"; and two, the charm of the older districts of San Francisco, Georgetown in Washington, DC, and Rittenhouse Square in Philadelphia; he identified the same qualities of small-scale, compact urbanism, vitality, and diversity that Jane Jacobs would limn two years later in her *Death and Life of Great American Cities*. "It seemed to me that urbanism's darkest hour was upon us: with the affluent escaping to the suburbs, poverty and dilapidation were coming to dominate most of downtown." He realized that "the paradox of contemporary urbanism" was the conflict and apparent incommensurability between these two desires, one for private, natural space, the other for urban sociality. Accordingly, he set himself "an architectural challenge: to invent a building type that provided the lifestyle of a house and garden, but that was compact enough to be constructed in the central city. This way, you could have your cake and eat it too." He gave his project the motto "For everyone a garden," and began to flesh out his thesis project by stacking Legos. With input from a year working with Louis Kahn in Philadelphia in 1963, he evolved a proposal for a community of living capsules made of precast concrete boxes, stacked to form a set of interlocking A-frames, much like a staggered set of ziggurats, linked by bridges. The thesis model and drawings caught the attention of the Expo 67 planners, and the young architect's utopian scheme improbably headed toward realization—though at 11 built stories instead of the originally intended 22, and much diminished in volume. Nevertheless, when the last piece was hoisted into place, a total of 354 prefabricated modules formed 158 apartments of varying sizes and layouts, each with at least one roof terrace for its private garden. Habitat 67 seemed to some like a science fiction version of a Mayan pyramid piled atop an island in the Saint Lawrence River, to others like a futurist re-creation of the mythical Italian hill town. Reyner Banham slyly noted a "picturesque disorder" in its jumbled quality and individualized gardens—implicitly comparing it to another utopian attempt to reconcile the city with the garden, the romantic garden suburbs. Habitat 67 was in many ways Frank Lloyd Wright's Usonian house multiplied and stacked—a hybrid form intended to reconcile the single-family house and the apartment building, the city and nature, and big and small. Ostensibly modular and prefabricated, it was, like Wright's Usonians, in reality too carefully customized and detailed, and thus too expensive, to replicate as the working man's paradise. Originally, Safdie's concept was for the boxes to be identical, but ultimately the architect had to accommodate different floor plans and the resulting asymmetrical loadings, as well as fire codes that led to each box weighing 70 to 90 tons. The cost reached $20 million—a fortune at the time.

Habitat 67 was a hit among the exposition-goers. Afterward it was sold to private owners and remains a prestigious address today. Many have asked, though, whether its success owed to interest in finding a new, more sustainable relationship between the city and the environment through technology and design, or whether it amounted to a futuristic, semivertical garden city. Safdie himself reflected on the study trip that launched him on his mission: "In retrospect, I had set out on this trip with preconceived ideas, feeling suburbia was bad—after all, the Mediterranean cities were my background. But my conclusion was new; I felt we had to find new forms of housing that would recreate, in a high density environment, the relationships and the amenities of the house and the village." Banham wondered if Safdie hadn't "merely backed up into his autobiography and built an image of his proto-Mediterranean prejudices, and that was the real reason for its success." Even if this were so, and Habitat 67 was not a leap forward but a pandering to mundane desires, these same desires underlay Metabolism, too, which also set out to satisfy them using technology. One might say in both cases that the problem remained the modern city's perennial one: how to be "in harmony with nature" and with traditional desires, and at the same time in harmony with technological progress and the extraordinary scale of modern populations. How could design give "everyone a garden" in megalopolises growing by millions every few years? Safdie would write, exactly three decades later: "The problem of scale is real: it is the result of fundamental changes in the statistical condition of humanity."

After Montreal's Expo 67, Metabolism and similar megastructural ideas seemed to appear everywhere—at least as audacious proposals on paper. Just a smattering were Frei Otto's High Rise project, with gardens spilling off its balconies, Walter Jonas's Intraraumstadt, Paolo Soleri's Arcology and Arcosanti, Manfredi Nicoletti's re-visions of New York, J. C. Bernard's Total City, Stanley Tigerman's Instant City, Claude Parent's Turbo City, Guy Rottier's Eco City, and Fuller and Sadao's Tetrahedron City. But what was actually built were megastructures tamed and scaled down to more consumable pieces: "habitats" that, like vehicles, could be deployed anywhere—perfect for an optimistic, universal future. Tange's broadcasting center in Kofu and Isozaki's library in Oita were both completed in 1968, and in 1969 Kurokawa built the Odakyu Drive-In Restaurant, a US-style diner on a hillside, reachable only by car, at the Hakone resort. It was to be the prototype for a gigantic space frame holding housing capsules, which was never built. Instead, it perversely but perfectly heralded the coming hegemony of US-style auto-dominated suburbia, with its environmentally unsustainable costs, to the entire planet.

The next Universal Exposition was held in Osaka, in 1970, in a transformed country—more confident since hosting the 1964 Tokyo Olympics, and in some way morally redeemed in the eyes of the world by becoming the second-largest economy on earth during the 1960s, based on its industrial and organizational prowess. Expo 70's theme was appropriately "Progress and Harmony for Humankind," and its chief planner was Kenzo Tange. With Isozaki and a team of engineers, Tange designed a "Big Roof" 30 meters high covering the festival plaza. Other Metabolists received major commissions: Kikutake designed the Expo Tower, Ekuan the street furniture and transportation, Otaka the transport center, Kawazoe the Mid-Air Exhibition. Kurokawa designed two corporate pavilions: the Toshiba pavilion, a biomorphic sculpture-object based on modular tetrahedrons that looked like a giant amoeba, and the Takara Beautilion, a prefabricated space frame of stacked cubes with capsules fit in to make showrooms, which took six days to build and plug in. The grounds were a triumph of Japanese technical wizardry and style, a miniature Utopia filled with consumer electronics and entertainment, including robots, the world's first videophone and first IMAX movie, and a flight simulator. As stunning as the theme buildings were, the corporate pavilions—one inflatable, one apparently levitating, one a flying saucer—were the most popular: Utopia was going to be a private, not a collective affair. In all, $2.9 billion was spent mounting the show, which over six months drew 64 million visitors.

The success of the Osaka Expo demonstrated the public's enthusiasm for the Metabolist aesthetic, at least in the limited, consumable objects on display (even if not acceptance of the full Metabolist program). The Metabolists basked in its glow. In 1972, Kurokawa's own capsule summer retreat, called House K, built on a hillside overlooking the sea in Karuizawa, Nagano prefecture, graced the front pages of magazines, and he built Metabolism's first and only residential building, the Nakagin Capsule Tower, in downtown Tokyo. His celebrity continued its rise; an interview with Kurokawa appeared in Japanese *Playboy* in February 1974. Outside Japan, the Metabolist program seized the pinnacle of the architectural firmament in 1971, when the team of Richard Rogers, Renzo Piano, and Gianfranco Franchini won the design competition for the enormous Pompidou Center in the heart of Paris. The spectacular and controversial design was straight out of an Archigram drawing, an inside-out building of diagonal steel frames, exterior escalators, brightly colored tubes, exposed and color-coded mechanical systems, floating interior layers, and radical transparency. In spite of determined consternation by many commentators, the building, and the lively street life it seemed to conjure on the Place Beaubourg in front of it, was an unequivocal smash with Parisians. It colorfully marked what Banham called "the institutionalization of megastructure," which gained momentum through the 1970s with a rash of more drab, concrete buildings, built mostly for universities, in the United States, Canada, the United Kingdom, Germany, and elsewhere.

Somewhere between the successes of Osaka's Expo 70 and the Pompidou Center the pendulum began to swing the other way. Hypothetical projects issuing from the architectural avant-garde had already reached absurd limits, epitomized by the 1969 proposal by the British duo Mike Mitchell and Dave Boutwell, published in *Domus* magazine, for Comprehensive City: a single building stretching in a straight line from San Francisco to New York, containing everything, leaving the rest of the North American continent untouched by man. And when, as if suddenly, the mainstream establishment bought into the concept of megastructures, the radicals headed for the exits. Cesar Pelli, who had proposed his share of monumental projects, including a building draped over the top of a mountain peak in Los Angeles, sniffed: "The only reason for megastructures is the ambition of architects." Growing criticism of the postwar era's faith in technology combined with doubt about the government-business partnership model that had held sway for decades. It was no longer plausible that the place of the individual in society could be safeguarded by gigantic, centrally designed projects—a capsule of one's own wasn't enough. Bigness was out; techno-optimism was replaced by techno-skepticism. Arata Isozaki, who had worked with the Metabolists but never officially joined them, opined that his colleagues had been "too optimistic. They really believed in technology, in mass production; they believed in systematic urban infrastructure and growth. The Metabolists had no skepticism toward their utopia." The problem of the environment, too, came to seem exacerbated by scale: reports and books rained down with titles like *The Silent Spring* (1962), *The Population Bomb* (1968), *The Limits to Growth* (1972), and *Small Is Beautiful* (1973). Civil rights and independence movements, student revolts, and calls for deep social and economic change didn't neglect architecture in their critiques. Banham's list of those who hated Megastructure revealed the broad dimensions of the cultural shift:

> For the flower-children, the dropouts of the desert communes, the urban guerrillas, the community activists, the politicized squatters, the Black Panthers, the middle-class amenitarians and the historical conservationists, the Marcusians, the art-school radicals and the participants in the street democracies of the événements de Mai, megastructure was an almost perfect symbol of liberal-capitalist oppression. It was condemned almost before it had a chance to happen.

The economic crises of the 1970s and the oil shocks of 1973–74 sealed the end of optimism. For Japan, the oil embargoes were a rude awakening: the Japanese economic miracle was dependent on the lifeblood of Middle East oil—a new vulnerability as frightening as tsunami, earthquakes, or bombs. In Japan and elsewhere, the techno-eco city program disappeared. But there were two notable exceptions to the counterculture's rejection of designers of megastructures. The first was the Italian-born architect Paolo Soleri, who had briefly been an apprentice with Frank Lloyd Wright at Taliesin West before drawing up his own monumentalist visions of "Arcology" in the Arizona desert. His vertical concrete supercities were really just Corbusian functionalism rendered as sci-fi space colonies, but Soleri became an unlikely guru of the back-to-the-land movement, partly owing to his interest in reducing man's urban footprint by building upward, and partly as a result of his failed attempt at building a real Utopia, Arcosanti in the Arizona desert, which was so unsuccessful that he and his disciples were forced to get by selling sand-cast ceramic bells. The second was Buckminster Fuller, whose concept of a resource-limited Spaceship Earth became a philosophical touchstone of the counterculture, and whose geodesic domes became home to tens of thousands across the world.

At the end of the day, Metabolist planning ideas had little visible effect in Japan, which continued to grow in its fragmented, haphazard way, in the furrows of its old landownership patterns and ever-rising land values—in its de facto traditional postwar way. Of the few constructed buildings that could be called Metabolist, weaknesses of the idiom were quickly discernible, beginning with the fact that concrete doesn't always age well. The most iconic, Kurokawa's Nakagin Capsule Building, still stands in central Tokyo, a magnet for architectural tourists, and what Koolhaas and Obrist called "a solitary time traveler from a thwarted future." More than four decades after it was built, a pair of Portuguese architects moved in for a time, reveling in the chance to live within the Metabolist dream:

"Nevertheless," they reflected, "it still feels like we are living somewhere in between a hotel and a scientific experiment."

Yet the demise of the megastructural era was greatly exaggerated. The austere aesthetic and the art of building with concrete and steel were carried forward and developed to exquisite levels in the architecture of Arata Isozaki, Fumihiko Maki, Kiyonori Kikutake, and other members of the 1960 group, as well as a younger generation of masters, including Tadao Ando and Kikutake's former associates Toyo Ito and Itsuko Hasegawa. Three of the six (Maki, Ando, and Ito) won the coveted Pritzker Prize in architecture. After 1973, in a fitting irony, many of the most prominent Metabolists fanned out to the developing world, especially to the modernizing oil states that had shown up Japan's limits, and to the newly independent African and Islamic Asian countries, where they designed and in some cases built stadiums, airports, universities, national assemblies, palaces, embassies, hotels, resorts, and city plans. In ambition, scale, technical sophistication, and groundbreaking systems of environmental control, many of these projects exceeded anything they had completed in Japan.

And actual megastructures proliferated in the 1970s, especially offshore oil drilling platforms, manifestly realizing the Metabolists' wildest dreams: huge cities of concrete and steel, built on artificial land or afloat on the oceans, self-sufficient in energy, electricity, and water desalination, largely prefabricated, adaptive, flexible—and, like Kikutake's description of his imagined Marine City of 1958, "moveable, autonomous, and climate-controlled." In 1973, the world's first concrete offshore rig was prefabricated and towed to the Ekofisk field in the North Sea off Norway, built by Phillips Petroleum and engineered by the Ove Arup Group, the same firm that engineered the Pompidou Center. Kikutake, leading a team of researchers at the University of Hawaii in the early 1970s, developed what became Aquapolis, a floating city that looked exactly like an oil platform, constructed in Hiroshima in 1975 and towed by tugboats 1,000 kilometers to Okinawa, where it served as the Japanese pavilion at Expo 75, held to mark Okinawa's return to Japan by the United States. Anchored offshore with a pier connecting it to the island, it measured 100 meters by 100 meters and contained a banquet room, offices, exhibition spaces, a post office, and housing for 40 people. With no oil of its own, Japan nonetheless pioneered the design and construction of the oil industry's expansion into the sea and provided it with some of the vision needed to embark on it.

Under the sea, too, new habitats proliferated: Conshelf I (Continental Shelf Station) was built in 1962 by Jacques-Yves Cousteau's group, funded by the French oil industry, and settled 10 meters deep off Marseilles. Conshelf, SEALAB, and Hydrolab followed. There was the German Helgoland Habitat in 1968, and Tektite, an offshoot of the Skylab program in space designed by General Electric and NASA, in 1969. And La Chalupa, an underwater habitat first immersed off of Costa Rica for research, was repurposed on the seafloor off Key Largo, Florida, as a hotel, where it hosted 10,000 overnight guests in the course of three decades. Domes and space frames, too, proliferated, from owner-built plywood geodesic homes to the soaring canopies of industrial facilities and airports—perhaps the ultimate self-contained, constantly changing structures—to the earthbound space colony experiment of Biosphere II, built in the Arizona desert from 1987 to 1991 and sealed off for two years with eight bionauts inside.

Through the 1970s, 1980s, and 1990s, habitats that would have been recognizable to the authors of *Metabolism 1960* spread across the sea, under it, into Antarctic ice fields, and into Earth's orbit.

Even if the connection between these experiments and the dreams of Kenzo Tange and his colleagues seems tenuous, those early visions of a technologically enabled, environmentally sustainable practice of city building had a direct line of influence on the architecture of the contemporary world in the twenty-first century. One case among many is provided by the career of the British architect Norman Foster, a working-class architecture student at the University of Manchester in the late 1950s who won a master's degree fellowship at Yale University in Connecticut for the academic year 1961–62, in part by dint of his skill at perspective drawing. At Yale, Foster was taught by the CIAM stalwart Paul Rudolph; the historian Vincent Scully; Serge Chermayeff, an eastern European emigré who had collaborated with Erich Mendelsohn in London; and the California modernists Craig Ellwood and Charles Eames, famous for their quasi-prefabricated Case Study house designs. During his fellowship year, Foster visited some buildings by Frank Lloyd Wright and Louis Kahn, and traveled to the West Coast, seeing examples of the fresh design ideas there and working for a time on the new University of California campus at Santa Cruz with several prominent West Coast architects, including William Wurster. He met another Brit in the Yale program, Richard Rogers, there with his wife, Su, who was equally excited by the open and optimistic American architecture culture. Back in the United Kingdom, they would start a London practice, working together until 1967, when they parted ways. Nineteen seventy-one was a significant year for each: Rogers won the Pompidou competition with his partners, and Foster met Buckminster Fuller, with whom he found numerous commonalities, including, Foster recalled, an attitude of "impatience and an irritation with the ordinary way of doing things." Foster began a long collaboration with Fuller and Sadao, Inc., working first on an innovative design for an underground theater at the University of Cambridge, then on a domed office building for the University of Oxford they called the Climatroffice, with layered workspaces set among lush interior landscaping—a reworking of the 1967 US Pavilion at Expo 67. Neither was built, but the Climatroffice concept transferred to Foster and Partners' project for an open-plan office building for the advertising firm of Willis Faber and Dumas, in Ipswich, England. Because of technical complexities, Foster couldn't deliver the pure concept on schedule, but the building that did result in 1975 used many of its ideas, including curving glass curtain walls and interior gardens. A new road beckoned.

Foster worked through the transition from drafting pencils and handheld calculators to computer-aided design. He would write at the beginning of the twenty-first century that "many of the 'green' ideas that we explored in early projects are only now being made possible by the new technologies at our disposal." Foster and his burgeoning firm went on to design a wide range of buildings over a half century: factories, offices, museums, concert halls, sports stadiums, rail stations, airports, private houses, and more, using the full panoply of megastructural materials and techniques: concrete, glass, and steel—able to be shaped into fluid curves through parametric computer modeling—diagonal bracing, steel trusses, space frames, exposed infrastructure, stayed masts, tension systems, prefabrication, and modular units. Most were explicitly designed around passive and active energy efficiency, natural lighting and ventilation, rainwater capture, high urban densities and transport linkages—all the ingredients of the ubiquitous rubric "sustainability." Foster talks often of "integration": "For me, the optimum design solution integrates social, technological, aesthetic, economic and environmental concerns." He echoes the Metabolist credo of adaptable buildings, writing: "Does the thinking behind their design anticipate needs that might not have been defined when they were created? Only time will tell, so we design buildings that are flexible and able to accommodate change." Indeed, a survey of his firm's enormous output shows Foster exploring and realizing the catalog of typologies proposed by an earlier generation of visionaries at the dawn of the 1960s. Domes appear frequently, such as in the Reichstag, the new German parliament in Berlin, or the Great Court of the British Museum, as do big roofs, the firm's form of choice for industrial buildings and airports. The firm's portfolio is thick with transportation hubs, from London's Stansted Airport, King's Cross and Saint Pancras railway stations, Canary Wharf tube station (with a glass dome), and Millennium Bridge, to sprawling airports in Beijing and Hong Kong—the latter of which would have pleased Kenzo Tange, having been built on "new land": a 100-meter-high island leveled and expanded by four times. Wind turbines, integrated power systems, and communication towers round out the infrastructural contributions. Metabolist visions of "organic" cities in the sky have found echoes in a series of Foster skyscrapers incorporating atriums and gardens, the most celebrated being the Commerzbank Headquarters building in Frankfurt, Germany, with its four-story atriums fitted with hanging gardens at intervals on the tower. The garden trope in particular, realized in small ways in most of his buildings and in spectacular form in the Wales National Botanic Garden's enormous dome and the Commerzbank tower's aerial atriums, pointed the way toward contemporary design's obsession with hanging horticulture, nurtured and protected in high-tech enclosures, onto buildings, as if to camouflage them, and, in the process, redeem the city by covering it with a green skin copied from nature.

Perhaps no building is better known as a Foster and Partners project and more identified with the firm's reputation for advanced technology and sustainability than the Swiss Re Headquarters, at 30 Saint Mary Axe, in London. Built on the former site of the Baltic Exchange, a building damaged by an IRA bomb, "the Gherkin," as it quickly became known after its plans were unveiled in 1997, was conceived to be "London's first ecological tall building," according to its designers' claims, with a double glass skin and computer-controlled operable windows to provide natural ventilation. The client, the reinsurance giant Swiss Re, is in the business of underwriting natural disasters and thus financially exposed to climate change; the company wanted an iconic building that would express forward thinking on energy use. "For us, sustainability makes excellent business sense," said one of its officers. The design would seem to have fulfilled in some measure, as far as one building could, the promise of Metabolism, by harnessing technology to make a new kind of human habitat that is adaptable, sustainable, and inspiring. The firm talked of a direct lineage to Fuller: early designs resembled the Climatroffice closely, and the finished structure, with its icosahedron frame and hollow transparency, looks like a rocket-shaped, computer-generated elongation of a geodesic dome.

And yet it raised as many questions as it answered. Can a building with a technologically "sustainable" design really mitigate the risks of climate change? Can architecture help us adapt to change, incrementally, without our altering the underlying economic system? Or does it simply underwrite and legitimate the business-as-usual model of unsustainable economic growth, built on irresponsible risk-taking? Problems with the Gherkin's operable windows forced the mixed-mode climate control system to be shut down or to operate sporadically, undermining its claims to efficiency, and perhaps making the glass structure even less efficient in reality than a conventional one would have been. Nevertheless, the insurance company sold the building in 2007 for $1.2 billion, for a profit of $400 million.

Undoubtedly, the Gherkin and other Foster projects spectacularly achieved the look and aspirations of a technological modernity seeking to be "in harmony with nature," making Foster the heir of Kenzo Tange. Foster has been duly honored with the trappings of architectural aristocracy: a knighthood in 1990, a life peerage—he is Baron Foster of Thames Bank—a Pritzker Prize, in 1999, and a career so dominant for so long as to leave few rivals. Twenty-first-century London, with its gleaming glass monuments to high culture and international finance, is itself virtually a monument to Foster and Partners' technical prowess, creative vision, aesthetics, philosophy, and business success. Foster's biggest achievement, and the unchallenged avatar of the megastructure, will be the two-mile-wide glass ring of the Apple computer campus in Cupertino, California, under construction as of this writing. For better or worse, the Apple building, designed for the world's largest consumer electronics company, completes a full circle back to Expo 70 in Osaka, when private corporations assumed leadership of an urban design mission that had its roots in collectivist Utopia. Megastructure was, it turned out, the perfect symbol of liberal capitalism. Foster's oeuvre, following in the footsteps of Fuller, Sadao, and the Metabolists, perfected a grammar of structure, system, and aspiration that has become, without question, the way the world now wants to build.

7. Habitats

Field Guide: Techno-Ecologies

Diagnostics

- Futuristic: constructed of steel, concrete, and glass, using innovative structural techniques, often experimental, including geodesic domes and other space frames, core-joint systems, or reinforced concrete. Often incorporating modules, capsules, and expandability.
- Integrated use: large structures incorporating many uses, either within a single building or in complexes, some approaching the size of city districts, or entire cities.
- Environmental orientation: incorporating "sustainability" measures such as climate control, water conservation, recycling, food production, or energy generation.

Examples

Japan

- Tokyo: Nakagin Capsule Tower (Kisho Kurokawa, 1972), Shizuoka Press and Broadcasting Center, Shimbashi (Kenzo Tange, 1967), Spiral Building (1985).
- Kofu: Yamanashi Communications Center (Kenzo Tange, 1968).
- Osaka: Umeda Sky City.

USA

- Arizona: Arcosanti (Paolo Soleri, ongoing).
- Orlando, Florida: Spaceship Earth, EPCOT Center, Disney World.
- Saint Louis: The Climatron (dome greenhouse), Missouri Botanical Garden (1960).

Canada

- Montreal: Habitat 67, Montreal Biosphere (former US Pavilion, Expo 67).

United Kingdom

- London: Thamesmead housing estate (a setting for the film *A Clockwork Orange)*, Alexandra Road housing, Spine Building, Bedford Way (Denys Lasdun, 1975), the Barbican Estate, Lloyd's of London (Richard Rogers, 1983); 30 Saint Mary Axe (the Gherkin).
- Norwich: University of East Anglia (plan by Denys Lasdun, 1966).

France

- Paris: Centre Georges Pompidou (Richard Rogers, Renzo Piano, and Gianfranco Franchini, 1971–).

Singapore

- Republic Polytechnic campus (2007).

Australia

- One Central Park, Sydney (Jean Nouvel, Patrick Blanc, 2014).

Variants

- Domed stadiums.
- Geodesic domes.
- Camping tents.

"Green" high-rises

(with plants prominently growing on or inside)

- Frankfurt, Germany: Commerzbank Headquarters (1991–97).
- New York: Hearst Tower, New York.
- Milan: Bosco Verticale.
- Sydney: One Central Park building (2014).

Living Walls

- Walls with "vertical gardens" pioneered by French designer Patrick Blanc.

Eco-cities

- Tianjin eco-city, China.
- Abu Dhabi Masdar City.
- Songdo, Korea.

Images:

Montreal Biosphere (1967), former US Pavilion, Expo 67, Montreal, Canada. Architects: Buckminster Fuller and Shoji Sadao. *Cédric THÉVENET*

Habitat 67 (1967), Montreal, Canada. Architect: Moshe Safdie.

Nakagin Capsule Tower (1972), Tokyo, Japan. Architect: Kisho Kurokawa. *Taxiarchos228 at the German language Wikipedia*

Shizuoka Press and Broadcasting Center (1967), Shimbashi, Tokyo, Japan. Architect: Kenzo Tange.

30 Saint Mary Axe, "The Gherkin" (2003), London, England. Architect: Foster Partners.

Reichstag, New German Parliament (1999), Berlin, Germany. Architect: Foster Partners.

Commerzbank Headquarters (1997), Frankfurt, Germany. Architect: Foster Partners.

Centre Georges Pompidou (1971–), Paris, France. Architects: Richard Rogers, Renzo Piano, and Gianfranco Franchini.

One Central Park (2014), Sydney, Australia. Architect: Atelier Jean Nouvel. Vertical garden by Patrick Blanc, irrigated with reclaimed sewage water. Cantilevered heliostats direct sunlight to shaded spaces.

Acknowledgments

This book would not have been possible without the help of many people, each of whom contributed a unique set of eyes and ears to the project: my editor Jennifer Barth at HarperCollins, ably seconded by Erin Wicks, Leah Carlson-Stanisic, Leslie Cohen, and Gregg Kulick; David Kuhn and Nicole Tourtelot at Kuhn Projects, Charles Donelan, Ann Ehringer, Otis Graham, Rowan Pelling, and Eddie Wang. I also want to acknowledge a debt of gratitude to an institution devoted to books, inside which this one gradually came into being: the Los Angeles Public Library, a thing simultaneously of bricks and mortar and web pages and the people who hold the two realms together, and its Central Library building in downtown LA, one of Bertram Grosvenor Goodhue's greatest achievements, still a breathing, inspiring masterpiece of urban vision.

Notes

Introduction

xi “the doctrine of salvation”: Quoted in Jane Jacobs, *The Death and Life of Great American Cities* (New York: Vintage Books, 1992), 113.

xi “In the end, I promise”: Lewis Mumford, *The Story of Utopias* (New York: Viking Press, 1962), 26.

Chapter 1: Castles

5 “drawing dream cities”: Charles Harris Whitaker, *Bertram Grosvenor Goodhue: Architect and Master of Many Arts* (New York: Press of the American Institute of Architects, 1925), 12–13.

5 So in 1884, at 15 years: Ibid, 16; also Richard Oliver, *Bertram Grosvenor Goodhue* (New York: Architectural History Foundation; Cambridge, MA: MIT Press, 1983), 5–6.

5 Setting aside his austere: Whitaker, *Bertram Grosvenor Goodhue*, 13.

7 “His pen and ink renderings”: Romy Wyllie, *Bertram Goodhue: His Life and Residential Architecture* (New York: W. W. Norton, 2007), 27.

7 The first portfolio, done in 1896: Oliver, *Bertram Grosvenor Goodhue*, 32.

9 “Below me in the now windless”: Ibid., 38.

10 The decade of the 1890s was the period: Ibid., 26.

11 Prodigious also was his vitality: Whitaker, *Bertram Grosvenor Goodhue*, 30.

11 “perched on a table, smoking”: Ibid, 31.

14 “morality of architecture”: John Ruskin, *The Stones of Venice* (New York: J. Wiley, 1877).

15 “The new society . . . has no prototype”: Stanley Schultz, *Constructing Urban Culture: American Cities and City Planning, 1800–1920* (Philadelphia: Temple University Press, 1989), 9.

15 "We are all a little wild here": Ibid.

15 Many experiments were utopian in intent: Ibid.

15 Some put faith in formal innovations: John Reps, *The Making of Urban America: A History of City Planning in the United States* (Princeton, NJ: Princeton University Press, 1965), 487–95.

16 "like so many paper soldiers": Schultz, *Constructing Urban Culture*; cf. Reps, *Making of Urban America*, 13.

17 Romanticism, as the German sociologist Georg Simmel noted: Chris Petit, "Bombing," in *Restless Cities*, ed. Mathew Beaumont and Gregory Dary (London: Verso, 2010), 29.

17 "despite their continuing mutual reinforcement": Kasia Boddy, "Potting," in Ibid., 214.

17 "smokestacks versus geraniums": Ibid., 218.

18 "a dry man": *The Project Gutenberg EBook of Great Expectations, by Charles Dickens*, www.gutenberg.org.

18 "The office is one thing": Ibid.

18 "By degrees, Wemmick got dryer": Ibid.

19 "They are wrapt, in this short passage": Quoted in Rachel Bowlby, "Commuting," in Beaumont and Dary, *Restless Cities*, 53.

20 "widely perceived to have never": Seth Lerer, *Children's Literature: A Reader's History, from Aesop to Harry Potter* (Chicago: University of Chicago Press, 2008), 257.

22 After fruitless meetings: Oliver, *Bertram Grosvenor Goodhue*, 22.

22 "So far, Mexico has not sunk" and other quotes: Bertram Grosvenor Goodhue, *Mexican Memories: The Record of a Slight Sojourn Below the Yellow Rio Grande* (New York, 1892), 9.

23 "Rapidly growing larger and larger": Ibid., 16.

23 "sombrero, zarape, a cool loose shirt": Ibid., 133.

23 "But you must make haste": Ibid., 135.

24 In late September 1894: Oliver, *Bertram Grosvenor Goodhue*, 25.

24 "it would be a mistake to try and define it": Quoted in Whitaker, *Bertram Grosvenor Goodhue*, 19.

24 "In France the cathedral stood": Oliver, *Bertram Grosvenor Goodhue*, 21.

25 Cram and Wentworth was busy: Ibid.

25 In 1897, partner Charles: Ibid., 12.

25 In December 1898, he returned: Ibid., 31.

27 "rather as a glamour than a memory": Whitaker, *Bertram Grosvenor Goodhue*, 42.

27 The first filmmakers came to the area: Kevin Starr, *Inventing the Dream: California Through the Progressive Era* (New York: Oxford University Press, 1985), 285.

30 "This superb creation": Whitaker. *Bertram Grosvenor Goodhue*, 45.

33 Hunt would go on: Kevin Starr, *Material Dreams: Southern California Through the 1920s* (New York: Oxford University Press, 1990), 195.

34 supervised by Carleton Winslow: Ibid., 282.

35 sets loomed over busy streets: Starr, *Inventing the Dream*, 338.

35 "The residential people of Los Angeles": Quoted in Starr, *Material Dreams*, 210.

36 The Los Angeles regional growth machine: Peter Hall, *Cities of Tomorrow: An Intellectual History of Urban Planning and Design in the Twentieth Century*, 3rd ed. (Oxford: Blackwell Publishing, 1988), 283.

Chapter 2: Monuments

41 "one of the foremost architects of the world": Thomas S. Hines, *Burnham of Chicago: Architect and Planner* (New York: Oxford University Press, 1974), xxiii.

42 His influence was matched by his size: Ibid., 234–36.

43 He was born in 1846: Ibid., 3.

43 "rarely studied and was always censured" : Ibid, 9.

43 "I shall try to become the greatest architect": Ibid., 9–12.

44 "There is a family tendency": Ibid., 13–14.

44 in Wight's office he met the partner he needed: Ibid., 17.

44 the two opened their own practice: Ibid., 18.

45 "his powerful personality was supreme": Ibid., xxiii.

45 "a dreamer, a man of fixed determination and strong will": Ibid., xxiv–xxv.

45 Burnham has often been compared: Ibid., 22.

46 Root acknowledged when he joked: Oliver Larkin, *Art and Life in America* (New York: Holt, Rinehart, and Winston, 1960 [1949]), 285.

46 In 1870, the country's population: Hines, *Burnham of Chicago*, 44–45.

47 As business and population grew: Ibid., 47.

47 To Sullivan, he embodied the spirit: Ibid., 24–25.

48 "An amazing cliff of brickwork": Ibid., 69.

49 It was to be a stupendous undertaking: Ibid., 73, 76, 78.

49 "abandon the conservatory aspect": Ibid., 78.

50 More and more Americans, made wealthy: Larkin, *Art and Life in America*, 293.

51 Saint-Gaudens expressed their self-consciousness: Hines, *Burnham of Chicago*, 90.

51 Using steel frames covered in wood: Ibid., 71, 89, 92, 112.

52 In the fair's six-month run: Ibid., 117.

52 "Chicago was the first expression": Ibid., 73.

53 journalist Henry Demarest Lloyd thought: Robert Fogelson, *Downtown: Its Rise and Fall, 1880–1950* (New Haven, CT: Yale University Press, 2001), 324.

54 Louis Sullivan later griped: Hines, *Burnham of Chicago*, xxvi, 120, 123.

54 "People are no longer ignorant": Ibid., 125.

54 He reaped much of the very public accolades: Ibid.

54 In 1896, he and his wife, Margaret: Ibid., 137.

54 "It was a perfect evening": Ibid.

55 "What had triumphed in 1896": Jackson Lears, *No Place of Grace: Antimodernism and the Transformation of American Culture, 1880–1920* (New York: Pantheon, 1981), 189.

55 At Burnham's death in 1912: Hines, *Burnham of Chicago*, 271.

57 "My own belief is that instead": Ibid., 143.

57 "How else can we refresh our minds": Ibid., 145.

60 The problem of the railroad was solved: Ibid., 153–54.

62 His 1907 master plan for Los Angeles: Jeremiah B. C. Axelrod, *Inventing Autopia: Dreams and Visions of the Modern Metropolis in Jazz Age Los Angeles* (Berkeley: University of California Press, 2009), 22.

62 His far more modest plan for Santa Barbara: Starr, *Material Dreams*, 263.

62 Robinson gave "the dream city" of the Chicago World's Fair: Charles Mulford Robinson, "Improvement in City Life: Aesthetic Progress," *Atlantic Monthly* 83 (June 1899). Online at: http://urbanplanning.library.cornell.edu/DOCS/robin_01.htm.

62 "the evil of the World's Fair triumph": Lewis Mumford, *Sticks and Stones: A Study of American Architecture and Civilization* (New York: Dover Publications, 1955), 130.

62 "At my feet lay a great city": Hines, *Burnham of Chicago*, 174.

63 "Municipal advance on aesthetic lines": Charles Mulford Robinson, "Improvement in City Life: Aesthetic Progress."

64 "a more agreeable city in which to live": Gray Brechin, *Imperial San Francisco: Urban Power, Earthly Ruin* (Berkeley: University of California Press, 1999), 178.

64 to "render the citizens cheerful, content, yielding, self-sacrificing": Ibid., 145.

64 where it was moved in 1925 from its original location: http://www.artandarchitecture-sf.com/california-volunteers.html.

65 "Manila may rightly hope to become an adequate": Hines, *Burnham of Chicago*, 210.

66 In 1901, Sir Aston Webb was commissioned: Thomas R. Metcalf, *An Imperial Vision: Indian Architecture and Britain's Raj* (Berkeley: University of California Press, 1989), 176.

69 He even talked, as Daniel Burnham did: Le Corbusier, *The City of Tomorrow and Its Planning*. Originally published as *Urbanisme*, 1929 (New York: Dover, 1987), 240–41.

Chapter 3: Slabs

79 "one unified watchmaking industry": J. K. Birksted, *Le Corbusier and the Occult* (Cambridge, MA: MIT Press, 2009), jacket copy.

80 Charles-Édouard was exposed early: Ibid., 120, 231.

80 the 1912 Villa Jeanneret-Perret influenced by John Ruskin: Ibid., 121.

81 To expand his education, Charles-Édouard set off on a trip: Charles Jencks, *Le Corbusier and the Tragic View of Architecture* (Cambridge, MA: Harvard University Press, 1973), 32.

82 "He is unloading ballast. That is how you rise": Birksted, *Le Corbusier and the Occult*, 10.

82 "LC is a pseudonym. LC creates architecture": Ibid.

82 Charles Jencks, wrote of him that: Jencks, *Le Corbusier and the Tragic View of Architecture*, 24, 54.

83 The architect's guiding principle was separation: Norma Evenson, *Le Corbusier: The Machine and the Grand Design* (New York: G. Braziller, 1970), 7.

84 "The lack of order to be found everywhere": Ibid., 9.

84 "It is the street of the pedestrian of a thousand years ago": Stanislaus von Moos, "From the 'City for 3 Million Inhabitants' to the 'Plan Voisin,'" in *Le Corbusier in Perspective*, ed. Peter Serenyi (Englewood Cliffs, NJ: Prentice-Hall, 1975), 135.

84 Louis Sullivan, the Chicago pioneer: Ibid., 125–38.

85 itself inspired by a long line of American visions: Jean-Louis Cohen, *The Future of Architecture, Since 1889* (New York: Phaidon Press, 2012), 89.

85 Hans Poelzig proposed a Y-shaped skyscraper: Mardges Bacon, *Le Corbusier in America: Travels in the Land of the Timid* (Cambridge, MA: MIT Press, 2001), 155.

85 Ludwig Hilberseimer drew his High Rise City: Cohen, *Future of Architecture*, 178–79.

85 "one experiences here the beneficent results": Evenson, *Le Corbusier*, 10.

86 Owen had called his brick quandrangles "moral quadrilaterals": Robert Fishman, *Urban Utopias in the 20th Century: Ebenezer Howard, Frank Lloyd Wright, Le Corbusier* (Cambridge, MA: MIT Press, 1977), 14.

86 Fourier's phalanstery contained theaters: Evenson, *Le Corbusier*, 32.

86 Le Corbusier would have known of the work of: Ibid., 13.

86 "Let us listen to the counsels of American": Reyner Banham, *A Concrete Atlantis: U.S. Industrial Building and European Modern Architecture, 1900–1925* (Cambridge, MA: MIT Press, 1986), 227.

87 "We must increase the open spaces": Le Corbusier, *The City of Tomorrow and Its Planning*, 167.

87 "A city made for speed is a city made for success.": Ibid. 179.

88 "was to prove to be one of the most influential": Birksted, *Le Corbusier and the Occult*, 304.

89 Le Corbusier's writing "had a hypnotic effect": Jencks, *Le Corbusier and the Tragic View of Architecture*, 64.

89 "Those hanging gardens of Semiramis": http://www.fondationlecorbusier.fr/corbuweb/morpheus.aspx?sysId=13&IrisObjectId=6159&sysLanguage=en-en&itemPos=2&itemCount=2&sysParentName=Home&sysParentId=65.

90 "those gloomy clefts of streets": Ibid.

90 "The idea of realizing [this urban vision] in the heart": Ibid.

90 "Therefore my settled opinion": Le Corbusier, *City of Tomorrow and Its Planning*, 96.

91 "1. Requisitioning of land": Le Corbusier, *The Radiant City* (New York: Orion Press, 1967), 148–52.

91 "organized slavery": Ibid.

92 as the scholar Sven Birksted has pointed out: Birksted, *Le Corbusier and the Occult*, 19, 21, 24–25; Cohen, *Future of Architecture*, 48, 57, 127.

93 Plans for Rio, Buenos Aires, and Montevideo show: Iñaki Ábalos and Juan Herreros, *Tower and Office: From Modernist Theory to Contemporary Practice* (Cambridge, MA: MIT Press, 2003), 16–19.

93 "His output of city plans is remarkable": Jencks, *Le Corbusier and the Tragic View*, 120.

94 In the fall of 1935, Le Corbusier spent: Bacon, *Le Corbusier in America*, 3.

95 Of New York, which he called "a barbarian city": Le Corbusier, *The City of Tomorrow and Its Planning*, 76.

95 "As for beauty, there is none at all": Le Corbusier, *City of Tomorrow*, 45, 76.

95 "Yes, the cancer is in good health": Evenson, *Le Corbusier*, 29.

95 "mighty storms, tornadoes, cataclysms . . . so utterly devoid of harmony": Ibid.

95 the *Herald Tribune* printed this headline: Bacon, *Le Corbusier in America*, 26.

95 He assiduously searched out power brokers: Ibid., 159.

96 Jacob Riis's view was typical: Fogelson, *Downtown: Its Rise and Fall, 1880–1950*, 325.

96 one Ohio official told Congress: Ibid., 345.

97 Meant as a jobs, not a housing program: Ibid., 340.

98 Tall concrete slab construction was already common: Samuel Zipp, *Manhattan Projects: The Rise and Fall of Urban Renewal in Cold War New York* (New York: Oxford University Press, 2010), 15.

99 "Residential, commercial, and industrial areas": *To New Horizons* (General Motors film, 1940), 19:30.

100 "a district which is not what it should be": Fogelson, *Downtown*, 348.

100 A Philadelphia judge opined: Ibid., 350.

101 prominent public housing advocate Catherine Bauer: John R. Short, *Alabaster Cities: Urban U.S. Since 1950* (Syracuse, NY: Syracuse University Press, 2006), 79–80.

101 "the business welfare state": Zipp, *Manhattan Projects*, 112.

101 The target was 18 square blocks: Ibid, 78.

102 Pittsburgh Equitable Life Assurance Society built: Ibid., 101.

102 "highest and best use": Short, *Alabaster Cities*, 20.

102 most American central business districts: Fogelson, *Downtown*, 271.

102 Critics scoffed, pointing out: Ibid., 278.

103 the Gowanus Parkway, which Robert Moses realized by: Ibid., 278.

104 "cities are created by and for traffic": Hilary Ballon and Kenneth T. Jackson, eds. *Robert Moses and the Modern City: The Transformation of New York* (New York: W. W. Norton, 2007), 124–25.

104 "must go right through cities": Ibid., 124.

104 And he cleared blighted districts: Ibid., 97.

104 "You can draw any kind": Robert Caro, *The Power Broker: Robert Moses and the Fall of New York* (New York: Knopf, 1974), 849.

104 Even with a federal "write-down" of two-thirds: Ballon and Jackson, eds. *Robert Moses and the Modern City*, 97.

105 One study following the first 500: Ibid., 102.

105 In Manhattan alone, his programs cleared: Ibid., 47–49.

105 Where the old land coverage had been: Ibid., 108.

105 Urban Renewal was really "negro removal": Short, *Alabaster Cities*, 21–22.

107 Marshall Berman and others have written, urbicide: Marshall Berman, "Falling," in Beaumont and Dary, *Restless Cities*, 128.

Chapter 4: Homesteads

115 At 8:30 p.m. on Monday, April 15, 1935: Patrick J. Meehan, ed., *Truth Against the World: Frank Lloyd Wright Speaks for an Organic Architecture* (Washington, DC: Preservation Press, 1992), 343.

116 Earlier that Monday: Ibid.

116 Aged 67 in 1935, he was best known: Fishman, *Urban Utopias in the 20th Century*, 94.

117 "the entrails of final enormity": Meehan, *Truth Against the World*, 345.

117 "Meantime, what hope of democracy left to us": Ibid.

117 "I have tried to grasp and concretely interpret": Ibid.

117 "The city will be nowhere, yet everywhere": Myron A. Marty, *Communities of Frank Lloyd Wright: Taliesin and Beyond* (DeKalb: Northern Illinois University Press, 2009), 159.

119 "Broadacre City is no mere back-to-the-land idea": Meehan, *Truth Against the World*, 345.

119 "The true center (the only centralization allowable)": Fishman, *Urban Utopias*, 129.

120 "the peculiar, inalienable right to live his own life": Ibid., 110.

120 "the great arterial": Frank Lloyd Wright, "Broadacre City: A New Community Plan" (1935), in *Frank Lloyd Wright: Essential Texts*, ed. Robert Twombly (New York: W. W. Norton, 2009), 261.

120 "Broadacre City is not merely the only democratic city": Alvin Rosenbaum, *Usonia: Frank Lloyd Wright's Design for America* (Washington, DC: Preservation Press, National Trust for Historic Preservation, 1993), 87.

120 "In Broadacres you will find not only a pattern": Meehan, *Truth Against the World*, 345.

120 Government would be "reduced to one": Wright, "Broadacre City," 260.

120 "economic, aesthetic and moral chaos": Meehan, *Truth Against the World*, 345.

121 "little farms, little homes for industry": Wright, "Broadacre City," 262.

121 "The waste motion, back and forth haul": Ibid., 260–61.

122 Howard's time in America, including on a Nebraska homestead: Hall, *Cities of Tomorrow*, 89.

122 Wright provided for a "community center": Fishman, *Urban Utopias*, 135.

122 Wright's conception went beyond other: Ibid., 123.

123 "The price of the major three to America": Twombly, *Frank Lloyd Wright: Essential Texts* (New York: W. W. Norton, 2009), 259.

123 "The actual horizon of the individual immeasurably widens": Ibid., 40–41.

124 "If he has the means": Ibid.

125 "the greatest architect of the 19th century": Rosenbaum, *Usonia*, 75.

125 with whom she had two children: Ibid., 39.

126 In November of that year, he met: Donald Leslie Johnson, *Frank Lloyd Wright Versus America: The 1930s* (Cambridge, MA: MIT Press, 1990), 6.

127 "Come with me, Ogilvanna": Bruce Brooks Pfeiffer and Gerald Nordland, eds., *Frank Lloyd Wright in the Realm of Ideas* (Carbondale: Southern Illinois University Press, 1988), 166.

127 Miriam, angry, succeeded in having them: Fishman, *Urban Utopias*, 118–19.

127 The press had a field day: Ibid., 119.

127 to work off the $43,000 outstanding debt: Johnson, *Frank Lloyd Wright Versus America*, 9.

127 "Not only do I intend to be the greatest": Ibid, frontispiece caption.

127 "After me, it will be 500 years before there is another": Johnson, *Frank Lloyd Wright Versus America*, 55.

128 These included meditations on cities: Pfeiffer and Nordland, *Frank Lloyd Wright in the Realm of Ideas*, 109.

128 In 1931, he embarked on what he termed: Johnson, *Frank Lloyd Wright Versus America*, 158–59.

128 In his Princeton lectures he'd called for: Pfeiffer and Nordland, *Frank Lloyd Wright in the Realm of Ideas*, 51.

129 "superfluous" but effective tools: Meryle Secrest, *Frank Lloyd Wright: A Biography* (Chicago: University of Chicago Press, 1998), 156.

129 Visitors to the Taliesin compound: Johnson, *Frank Lloyd Wright Versus America*, 61, 63, 64.

129 The German modernist architect Ludwig Mies van der Rohe: Ibid., 177.

129 His first plans for communities: Ibid., 141, 147.

129 He had published an early version of the idea: "Plan by Frank Lloyd Wright," in *City Residential Development: Studies in Planning* (University of Chicago Press, May 1916). Marty, *Communities of Frank Lloyd Wright*, 167.

129 having traveled across the United States by car: Marty, *Communities of Frank Lloyd Wright*, 166.

129 In 1932, he had considered problems of: Johnson, *Frank Lloyd Wright Versus America*, 129.

130 "Modern transportation may scatter the city": Rosenbaum, *Usonia*, 65.

130 to convince Tom Maloney of New York to give $1,000: Johnson, *Frank Lloyd Wright Versus America*, 110.

130 Edgar Kaufmann, Sr., the Pittsburgh department store owner: Marty, *Communities of Frank Lloyd Wright*, 161–62.

130 In the final plan, finished in late 1934: Johnson, *Frank Lloyd Wright Versus America*, 112.

130 Work began on the Broadacre City model: Ibid., 115.

130 The fellows spent hundreds of hours: Marty, *Communities of Frank Lloyd Wright*, 159.

130 "We live in this future city. Speed in the shady lanes": Ibid, 163.

131 "Like the prophet of old reminding the children": Pfeiffer and Nordland, *Frank Lloyd Wright in the Realm of Ideas*, 150.

132 "There is no such thing as creative except": Ibid., 89.

132 The Chicago where he spent his early: Rosenbaum, *Usonia*, 26–27.

133 "The next America would be a collectivist democracy": Ibid., 99.

133 The influential writer Lewis Mumford: Ibid, 71, 73.

134 Ford had built his first car in 1893: Ibid., 48.

134 "I am a farmer. . . . I want to see every acre": Hall, *Cities of Tomorrow*, 275.

134 "Plainly . . . the ultimate solution will be the abolition of the City": Johnson, *Frank Lloyd Wright Versus America*, 134.

134 where 250 model Ford Homes were put up: Joseph Oldenburg, "Ford Homes Historic District History," http://www.fordhomes.org/fhhd_history.pdf.

135 At Florence, they discussed Ford's plan: Rosenbaum, *Usonia*, 55–57.

135 Wright echoed this title in his own later article: Johnson, *Frank Lloyd Wright Versus America*, 137.

135 "Even that concentration for utilitarian purposes": Ibid.

136 "a valley inhabited by happy people": Rosenbaum, *Usonia*, 92.

136 the TVA's first dam, Norris Dam, was built: Ibid., 124.

137 "finest city in the world . . . communicated with the Suburban Division": Rosenbaum, *Usonia*, 143.

137 The model was seen by 40,000 people: Marty, *Communities of Frank Lloyd Wright*, 165.

137 two long pieces in the *Washington Post*: Rosenbaum, *Usonia*, 143.

137 "No planning proposal has ever had as much exposure": Marty, *Communities of Frank Lloyd Wright*, 165.

138 "I am not guilty of offering a plan for immediate use": Fishman, *Urban Utopias*, 95.

138 "a naive concoction of adolescent idealism": (Stephen Alexander) Rosenbaum, *Usonia*, 120.

138 "would require the abrogation of the Constitution": Pfeiffer and Nordland, *Frank Lloyd Wright in the Realm of Ideas*, 159–60.

138 Wright had had experience with building smaller houses: Rosenbaum, *Usonia*, 133.

139 the first prefab community in the United States: Ibid., 180.

140 "He simply did not know how to do prefabrication": Ibid., 183.

140 Another of Wright's Broadacre inventions, the Roadside Market: Richard Longstreth, *The Drive-in, the Supermarket, and the Transformation of Commercial Space in Los Angeles, 1914–1941* (Cambridge, MA: MIT Press, 1999), 134.

141 In Southern California, by 1941 nearly half: Greg Hise, *Magnetic Los Angeles: Planning the Twentieth Century Metropolis* (Baltimore: Johns Hopkins University Press, 1997), 129.

141 By 1943, there were 400,000 defense workers: "The Economic Development of Southern California, 1920–1976," in *The Aerospace Industry as the Primary Factor in the Industrial Development of Southern California: The Instability of the Aerospace Industry, and the Effects of the Region's Dependence on It*, vol. 1 (Los Angeles: City of Los Angeles, Office of the Mayor: June 1976), 5–6.

141 like those offered by Pacific Ready Cut Homes: Becky Nicolaides, "'Where the Working Man Is Welcomed': Working-Class Suburbs in Los Angeles, 1900–1940," in *Looking for Los Angeles: Architecture, Film, Photography, and the Urban Landscape*, ed. Charles Salas and Michael Roth (Los Angeles: Getty Research Institute, 2001), 76, 77.

141 The resulting landscape was a mostly unplanned: Ibid., 78.

141 But as the number of war workers grew: Hise, *Magnetic Los Angeles*, 129.

142 Critical to their efforts were a raft of new: Ibid., 119, 134.

142 In 1940, Douglas Aircraft broke ground on a tract: D. J. Waldie, *Holy Land: A Suburban Memoir* (New York: St. Martin's Press, 1996), 25, 45.

142 The builders used a continuous-flow assembly line: Hise, *Magnetic Los Angeles*, 121, 137–140, 142–47.

142 The postwar Los Angeles that emerged: Ibid., 132, 190.

144 "Their overriding purpose": Mike Davis, *City of Quartz: Excavating the Future in Los Angeles* (New York: Vintage Books, 1992), 161.

144 But this supposed homogeneity had been eroded: James R. Wilburn, "Social and Economic Aspects of the Aircraft Industry in Metropolitan Los Angeles During World War II" (PhD dissertation, University of California, Los Angeles, 1971), 184–87.

144 "100% American Family Community": Waldie, *Holy Land*, 160.

144 The decision to incorporate was swung by: Historical sources at City of Lakewood website: http://www.lakewoodcity.org/about_lakewood/community/default.asp.

145 As further inducement, in 1956: Davis, *City of Quartz*, 166.

145 The Lakewood Plan spread like wildfire: Charles F Waite, "Incorporation Fever: Hysteria or Salvation: An Excerpt from a Report on Incorporation and Annexation in Los Angeles County, Prepared for the Falk Foundation" (Los Angeles: Los Angeles Bureau, Copley Newspapers, July 1952).

146 By the 1990s, there were 16,000 HOAs in California: Davis, *City of Quartz*, 160–65.

146 "diffuse, de-centered, without clear boundaries": John Dutton, *New American Urbanism: Re-Forming the Suburban Metropolis* (Mila: Skira Editore, 2000), 17.

147 "That's the big, big failure of Frank Lloyd Wright": Marty, *Communities of Frank Lloyd Wright*, 167.

Chapter 5: Corals

154 "Moses had more power over the physical development": Hilary Ballon, "Robert Moses and Urban Renewal: The Title I Program," in *Robert Moses and the Modern City: The Transformation of New York*, ed. Hilary Ballon and Kenneth Jackson (New York: W. W. Norton, 2007), 97.

154 "enabled him to accomplish a vast amount": Robert Fishman, "Revolt of the Urbs: Robert Moses and His Critics," in Ballon and Jackson, *Robert Moses and the Modern City*, 122.

154 "a few rich golfers": Ibid.

155 not only did his plans include a roadway: Ibid., 243.

155 "The democratic way is to allow the people": Ibid., 94.

156 "Project That Would Put New Roads in Washington Square Park": Ibid., 124.

157 "It is our view that any serious tampering with Washington Square Park": Ibid., 126.

157 "I don't care how those people feel": Ibid.

158 He damned what he called the dominance of traffic engineers: Ibid.

158 who lived with her husband and three children: https://en.wikipedia.org/wiki/Jane_Jacobs.

158 "He stood up there gripping the railing": Ballon and Jackson, *Robert Moses and the Modern City*, 125.

159 "It is no surprise that, at long last, rebellion is brewing in America": Ibid., 124.

160 in 1947 the couple bought a modest three-story: http://untappedcities.com/2010/09/28/jane-jacobs-house-at–555-hudson-street/.

161 "This book is an attack": All quotes from Jane Jacobs, *The Death and Life of Great American Cities* (New York: Vintage Books, 1992).

170 a 1962 *New Yorker* magazine article: http://www.placematters.net/node/1867.

170 The book would eventually be translated into six: http://en.wikipedia.org/wiki/The_Death_and_Life_of_Great_American_Cities.

172 The condominium was one of the most significant developments: https://en.wikipedia.org/wiki/Condominium; http://www.condopedia.com/wiki/Timeline_of_Condo_History; http://www.deseretnews.com/article/765615708/Father-of-Modern-Condominiums-will-never-live-in-one.html.

174 "A large or a small city can only be reorganized": Krier, "The City Within the City", in Demetri Porphyrios, ed., *Léon Krier: Houses, Palaces, Cities* (London: Academy Publications, 1984).

175 "After the crimes committed against the cities": *A + U*, Tokyo, Special Issue, November 1977, 69–152. Reprinted in *Architectural Design* 54 (July/August 1984): 70–105. Also in Porphyrios, ed., *Léon Krier*; http://zeta.math.utsa.edu/~yxk833/KRIER/city.html.

175 "One day I went to a lecture by Leon Krier": http://zeta.math.utsa.edu/~yxk833/KRIER/index.html.

177 "Somewhere along the way, traditional towns": Andres Duany, Elizabeth Plater-Zyberk, and Jeff Speck, *Suburban Nation: The Rise of Sprawl and the Decline of the American Dream* (New York: North Point Press, 2000), xi.

177 The "solution" they wrote, "is not removing cars": Ibid., 160.

177 "following six fundamental rules that distinguish it from sprawl": Ibid., 15.

178 "the key to active street life": Ibid., 156.

179 By guaranteeing "a consistent streetscape": Ibid., 177.

181 at 18 to 24 units per acre: Dutton, *New American Urbanism*, 51–52.

181 "a campaign to rescue the landscape": Howard Kunstler, *The Geography of Nowhere* (New York: Simon and Schuster, 1993).

182 CIAM, in Duany's phrase, "can be credited or blamed": Duany, Plater-Zyberk, and Speck, *Suburban Nation*, 253.

183 "add up to a high quality of life": Principle 10 of the New Urbanism: http://www.newurbanism.org/newurbanism/principles.html.

184 working name for the project was Dream City: Robert H. Kargon and Arthur P. Molella, *Invented Edens: Techno-Cities of the Twentieth Century* (Cambridge, MA: MIT Press, 2008), 135.

184 the median home price in Celebration was two times higher: Andrew Ross, *Celebration Chronicles: Life, Liberty, and the Pursuit of Property Value in Disney's New Town* (New York: Ballantine Books, 2011), 32.

185 "the sleaze road of all times": Ibid., 277.

185 "We are prepared to sacrifice architecture on the altar of urbanism": Duany, Plater-Zyberk, and Speck, *Suburban Nation*, 210–11.

185 "nonideological" and "the Mazda Miata, a car that looks": Ibid., 254.

186 "It is hard enough convincing suburbanites": Ibid., 210.

186 "as camouflage for subversive density, difference, and mixed use": Dutton, *New American Urbanism*, 67.

186 "the situation is so critical that Andres Duany and I": http://www.planetizen.com/node/32.

186 Robert Fishman has dubbed the "fifth migration": Duany, Plater-Zyberk, and Speck, *Suburban Nation*, 230.

187 her former home at 555 Hudson Street: http://untappedcities.com/2010/09/28/jane-jacobs-house-at–555-hudson-street/.

187 "safeguards against incongruity": Dutton, *New American Urbanism*, 79.

Chapter 6: Malls

199 In 1786, the duke invested in stone colonnades: http://www.histoire-image.org/site/oeuvre/analyse.php?i=684.

199 it offered new marketing possibilities for luxury: Johann Friedrich Geist, *Arcades, the History of a Building Type* (Cambridge, MA: MIT Press, 1983), 60.

199 In 1791, the Passage Feydeau set: Ibid., 4.

200 "These arcades, a recent invention of industrial luxury": Walter Benjamin, "The Paris of the Second Empire in Baudelaire," quoted in David S. Ferris, *The Cambridge Introduction to Walter Benjamin* (Cambridge: Cambridge University Press, 2008), 36–37.

200 By 1890, the United States could boast the largest: James J. Farrell, *One Nation Under Goods: Malls and the Seductions of American Shopping* (Washington, DC: Smithsonian Books, 2003), 5.

201 "The great poem of display chants its many-colored": Ferris, *Cambridge Introduction to Walter Benjamin*, 117.

201 the arcades were "the scene of the first gas lighting" . . . "its chronicler and its philosopher": Ibid., 36–37.

201 Benjamin saw in the arcades the apotheosis . . . where capitalism connected to the world of dreams: Ibid., 116.

202 From 1816 to 1840 many were built: Geist, *Arcades*, 49.

202 In 1851, as the centerpiece of London's Great Exhibition: Louise Wyman, "Crystal Palace," in Chuihua Judy Chung and Sze Tsung Leong, eds., *Harvard Design School Guide to Shopping* (Köln: Taschen, 2001), 236.

203 "We freely admit, that we are lost in admiration": Ibid.

204 "The total effect is magical, I had almost said intoxicating.": Ibid., 238.

204 In the wake of Paxton's achievement: Ibid., 33.

205 "a cathedral of modern trade, light yet solid": Mark Moss, *Shopping as an Entertainment Experience* (New York: Lexington Books, 2007), xx.

206 "the most monumental commercial structure ever": Hines, *Burnham of Chicago*, 303.

210 In 1954, Northland Center, designed by: Farrell, *One Nation Under Goods*, 8.

210 Gruen, an Austrian Jew originally named Grünbaum: Victor Gruen, *The Heart of Our Cities: The Urban Crisis; Diagnosis and Cure* (New York, Simon and Schuster, 1964), 10.

211 After he relocated to Beverly Hills, California: M. Jeffrey Hardwick, *Mall Maker: Victor Gruen, Architect of an American Dream* (Philadelphia: University of Pennsylvania Press, 2004), 35.

212 It would also include a series of functions: Ibid., 72–86.

212 Gruen completely enclosed the mall: Frances Anderton et al., *You Are Here: The Jerde Partnership International* (London: Phaidon, 1999), 46.

212 "In providing a year-round climate of 'eternal spring'": Chung and Leong, *Harvard Design School Guide to Shopping*, 116.

213 "recipe for the ideal shopping center": Ibid., photo caption.

214 "the Gruen Transfer": Farrell, *One Nation Under Goods*, 27.

214 Tax increment financing: Ibid 219–220.

214 Federal financial rules capping the amount: Robert Steuteville, "Restoring the Lifeblood to Main Street," http://bettercities.net/article/restoring-lifeblood-main-street-21194.

215 said to have at one time owned one-tenth: http://en.wikipedia.org/wiki/Edward_J._DeBartolo,_Sr.

215 "the best investment known to man": Margaret Crawford, "The Architect and the Mall," in Anderton et al., *You Are Here*, 45.

215 Victor Gruen's malls were in demand: http://mall-hall-of-fame.blogspot.com/2008_05_01_archive.html.

215 "boring amorphous conglomeration which I term 'anti-city'": Gruen, *Heart of Our Cities*, 65.

215 "Shopping centers have taken on the characteristics of": Chung and Leong, *Harvard Design School Guide to Shopping*, 384.

215 comparing them to the "community life that the ancient Greek Agora": Farrell, *One Nation Under Goods*, 9–10.

215 "a social, cultural and recreational crystallization": Gruen, *Heart of Our Cities*, 191.

216 "In the sound city there must be a balance": Ibid., 28.

216 "three qualities or characteristics that make a city": Ibid.

216 he even offered a romantic quote from: Ibid., 19.

217 "The plan by Victor Gruen": Ibid., 220

218 In 1954, suburban malls surpassed metro: Farrell, *One Nation Under Goods*, 11.

218 In 1968, Victor Gruen retired, embittered: Hardwick, *Mall Maker*, 216–17.

218 Gruen and Associates, eventually built 45: Chung and Leong, *Harvard Design School Guide to Shopping*, 742.

218 In a 1978 London speech: Victor Gruen, "The Sad Story of Shopping Centers," *Town and Country Planning* 46 (1978): 350–52.

218 "the most magic plan and the largest": Chung and Leong, *Harvard Design School Guide to Shopping*, 746.

219 He got his own chance to do an urban project: Joshua Olsen, *Better Places, Better Lives: A Biography of James Rouse* (Washington, DC: Urban Land Institute, 2003), 242.

219 "I wouldn't put a penny downtown": Ibid., 248–49.

220 "The American dream for millions and millions of young Americans": Ibid., 268.

220 "Shopping is increasingly entertainment": Moss, *Shopping as an Entertainment Experience*, 111.

221 "Profit is the thing that hauls dreams into focus": Farrell, *One Nation Under Goods*, 263.

222 The time shoppers spent in malls declined: Moss, *Shopping as an Entertainment Experience*, 59.

223 "Retail is the bottom of the bucket": Anderton et al., *You Are Here*, 46.

224 "a deliberate urban script, a conscious creation": Jon Jerde, *The Jon Jerde Partnership International: Visceral Reality* (Milan: l'Arca Edizioni Spa, 1998), 8.

224 Born in 1940 in Alton, Illinois: Cathie Gandel, *Jon Jerde in Japan: Designing the Spaces Between* (Glendale, CA: Balcony Press, 2000), 21.

224 "a wonderful warmth and sense of belonging": Anderton et al., *You Are Here*, 17.

224 During a yearlong travel fellowship: Ibid., 18.

225 "reinvention of communal experience": Jerde, *Jon Jerde Partnership International*, 9.

225 In Southern California, he knew of older examples: Marcy Goodwin, "One Hand Clapping," in *John Jerde: Redesigning the City*, exhibition catalog, organized by the San Diego Art Center, May 2–September 7, 1986, 4–5.

225 "scripting the city . . . to create urban theater": Ibid., 9.

225 "The shopping center is a pretty pathetic venue": Anderton et al., *You Are Here*, 18.

226 "In Horton Plaza's 40 acres we tried": Goodwin, "One Hand Clapping," 8.

226 a way to distill the "personality or persona": Jerde, *Jon Jerde Partnership International*, 10.

227 "a spirit afloat in Southern California": Ibid., 13.

228 25 million visitors in its first year: Anderton et al., *You Are Here*, 9.

228 "rejuvenate the complex using urban values": Jerde, *Jon Jerde Partnership International*, 43.

229 it was "going to be like going to Disneyland": Moss, *Shopping as an Entertainment Experience*, 19.

229 "What they wanted was four malls bolted": Chung and Leong, *Harvard Design School Guide to Shopping*, 534.

229 a claimed 40 million visitors each year: Moss, *Shopping as an Entertainment Experience*, 49.

229 Reilly's Law of Retail Gravitation—which states that: Chung and Leong, *Harvard Design School Guide to Shopping*, 532.

229 "Communal experience is a designable event": Jerde, *Jon Jerde Partnership International*, 9.

230 "Our stuff isn't supposed to be visual": Anderton et al., *You Are Here*, 129.

230 "He throws large amounts of architectural matter": Chung and Leong, *Harvard Design School Guide to Shopping*, 403.

231 "retail architecture is below the trash can": Gandel, *Jon Jerde in Japan*, 18.

232 Eighty thousand people came to see the spectacle: http://jerde.com/featured/place29.html.

233 "What we've figured out is that place": Jerde, *Jon Jerde Partnership International*, 11.

233 "We are like psychoanalysts, uncovering the dreams": Anderton et al., *You Are Here*, 176.

235 "a strange new animal": Chung and Leong, *Harvard Design School Guide to Shopping*, 534.

235 "shopping's effectiveness in generating constant activity": Sze Tsung Leong, "Mobility" in Chung and Leong, *Harvard Design School Guide to Shopping*, 477.

235 "over one billion people a year visit our projects": Jerde, *Jon Jerde Partnership International*, 10.

236 Craig Hodgetts asked whether "Jerde's artificial cosmos may": Craig Hodgetts, "And Tomorrow . . . The World?" in Anderton et al., *You Are Here*, 190.

Chapter 7: Habitats

245 another proposal Fuller and Sadao unveiled: "Project for Floating Cloud Structures (Cloud Nine)," ca. (1960). Black-and-white photograph mounted on board. 15 ⅞ x 19 ¾ in. (40.3 x 50.2 cm.) https://cup2013.wordpress.com/tag/shoji-sadao/.

247 "Into architecture," he would later recall: David B. Stewart, *The Making of a Modern Japanese Architecture: 1868 to the Present* (Tokyo: New York: Kodansha International, 1987), 170.

247 From 1938 on, Tange worked on: Ibid., 164.

248 In 1949, Tange's reconstruction plan: Kenzo Tange and Udo Kulturmann, *Kenzo Tange: Architecture and Urban Design* (New York: Praeger, 1970), 17–27.

248 Tange was invited to accompany: Stewart, *Making of a Modern Japanese Architecture*, 175.

251 "Like sea plants, an expandable chain of alternating balls": Rem Koolhaas and Han Obrist, *Project Japan: Metabolism Talks* (Koln: Taschen, 2011), 137.

252 "Tokyo is expanding but there is no more land": Reyner Banham, *Megastructure: Urban Futures of the Recent Past* (London: Thames and Hudson, 1976), 47.

252 Styling itself Team X: Stewart, *Making of a Modern Japanese Architecture*, 179.

253 "a kaleidoscopic inventory": Koolhaas and Obrist, *Project Japan*, 13.

254 "Architecture, which hitherto was inseparable from the earth": Ibid., 341.

254 "to understand the shift": Ibid., 19

255 Attending the World Design Conference: Stewart, *Making of a Modern Japanese Architecture*, 179.

259 It was followed by "Megastructure Model Kit": Banham, *Megastructure*, 98.

259 More "fun" structures were imagined: Ibid., 100–101.

260 "Three-quarters of our planet Earth": R. Buckminster Fuller, *Critical Path* (New York: St. Martin's Press, 1981), 333.

260 "While the building of such floating clouds: Ibid., 337.

261 "to call it a 'collapsed and rusting Eiffel Tower'": Banham, *Megastructure*.

261 a 5/8 geodesic dome, 200 feet high and 250 feet in diameter: http://expo67.ncf.ca.

262 A catalog of futuristic transport options: Banham, *Megastructure*, 116.

263 "It seemed to me that urbanism's darkest hour was upon us": Moshe Safdie, *The City After the Automobile: An Architect's Vision* (New York: Basic Books, 1997), x.

263 "an architectural challenge: to invent a building type": Ibid.

263 "picturesque disorder": Banham, *Megastructure*, 108.

264 "In retrospect, I had set out on this trip": Ibid., 111.

264 "merely backed up into his autobiography": Ibid.

265 "The problem of scale is real": Safdie, *City After the Automobile*, 90.

266 Other Metabolists received major commissions: Koolhaas, *Project Japan*, 507.

266 $2.9 billion was spent mounting the show: Ibid.

267 "the institutionalization of megastructure": Banham, *Megastructure*, 130.

267 epitomized by the 1969 proposal by the British duo: Ibid., 196.

267 "The only reason for megastructures": Ibid., 208–9.

268 his colleagues had been "too optimistic. They really believed in technology": Koolhaas and Obrist, *Project Japan*, 25.

268 "For the flower-children, the dropouts": Ibid., 209.

269 "a solitary time traveler from a thwarted future": Ibid., 389.

270 "Nevertheless, they reflected, "it still feels": http://www.domusweb.it/en/architecture/2013/05/29/the_metabolist_routine.html.

272 "impatience and an irritation with the ordinary way of doing things": *Foster Catalogue 2001* (Munich; New York: Prestel, 2001), 29.

273 "many of the 'green' ideas": Ibid., 6.

273 "For me, the optimum design solution": Ibid.

274 "Does the thinking": Ibid., 6.

275 "London's first ecological tall building": http://www.archdaily.com/447205/the-gherkin-how-london-s-famous-tower-leveraged-risk-and-became-an-icon-part–2/.

275 "For us, sustainability makes excellent business sense," said one of its officers: Ibid.

276 the insurance company sold the building in 2007: Fiona Walsh, "Gherkin Sold for £600m," *Guardian*, February 5, 2007.

Index

Page numbers in *italics* refer to illustrations.

About the Author

Wade Graham is a Los Angeles–based garden designer, historian, and writer whose work on the environment, landscape, urbanism, and the arts has appeared in *The New Yorker*, *Harper's*, *Los Angeles Times*, *Outside*, and other publications. An adjunct professor of public policy at Pepperdine University, he is the author of *American Eden: From Monticello to Central Park to Our Backyards: What Our Gardens Tell Us About Who We Are*.

About the Author